The Soul of
BASEBALL

The Soul of

BASEBALL

**A Road Trip Through
Buck O'Neil's America**

JOE POSNANSKI

WM

WILLIAM MORROW
An Imprint of HarperCollinsPublishers

HarperCollins books may be purchased for educational, business, or sales promotional use. For information please write: Special Markets Department, HarperCollins Publishers, 10 East 53rd Street, New York, NY 10022.

Designed by Chris Welch

Library of Congress Cataloging-in-Publication Data has been applied for.

ISBN: 978-0-06-085403-4
ISBN-10: 0-06-085403-0

07   08   09   10   11   WBC/RRD   10   9   8   7   6

FOR MARGO

CONTENTS

SUMMER

AUTUMN

WINTER (TAKE 2)

⋆ *The Soul of* ⋆
BASEBALL
⋆

WARMING UP A RIFF

We were in Houston in springtime. We sat in a ballpark under a sun so hot the seats melted beneath us. There is something honest about Houston heat—it comes at you straight. It does not drain you like the Washington humidity or try and trick you like dry heat in Phoenix. In Houston, the heat punches you in the gut again and again. Buck O'Neil was wilting.

"I'm ready to go back to the hotel whenever you are," Buck said to no one in particular, but mostly to me. We were at a ballgame. The baseball season had just begun. Before our road trip ended, Buck and I would go to many ballgames together. We would spend a full year, winter through winter, rushing to Buck's next appearance, ballpark to hotel to autograph session to school to hotel to museum and back to ballpark—thirty thousand air miles and another few thousand more on the ground. We traveled around America. Buck talked about baseball.

In time, I would grow accustomed to Buck's moods, his habits, his style, the way he wore his hat, the way he sipped his tea, the way he walked and talked, and the way he dressed. Buck splashed color. He wore bright crayon shades: royal purple, robin's-egg blue, olive green, midnight blue, and lemon yellow. He wore pinstripe white suits, orange on orange, and shoes that perfectly matched the color of his pants. He never wore gray.

In time, I would grow accustomed to Buck's boundless joy. That joy went with him everywhere. Every day, Buck hugged strangers, invented nicknames, told jokes, and shared stories. He sang out loud and danced happily. He threw baseballs to kids and asked adults to tell him about their parents, and he kept signing autographs long after his hand started to shake. I heard him leave an inspiring and heartfelt two-minute phone message for a person he had never met. I saw him take a child by the hand during a class, another child grabbed her hand, and another child grabbed his, until a human chain had formed, and together they curled and coiled between the desks of the classroom, a Chinese dragon dance, and they all laughed happily. I saw Buck pose for a thousand photographs with a thousand different people, and it never bothered him when the amateur photographer fumbled around, trying all at once to focus an automatic camera, frame the shot like Scorsese, and make the camera's flash pop at two on a sunny afternoon. Buck kept his arm wrapped tight around the women standing next to him.

"Take your time," he always said. "I like this." Always.

"Man, it's hot in Houston," Buck said, and he launched into a story about one of his protégés, Ernie Banks, the most popular baseball player ever to take Wrigley Field on the North Side of Chicago. Banks played baseball with unbridled joy. They called him "Mr. Cub." Funny thing, when Banks first signed with the Kansas City Monarchs—Banks was nineteen then, it was 1950—he was a shy kid from Texas. He sat in the back of the team bus and hardly spoke—"Shy beyond words," Buck called him. Buck was the manager of that Monarchs team, and he would say to Banks, as he said to all his players, "Be alive, man! You gotta love this game to play it."

Ernie Banks embraced those words. He opened up. His personality emerged. "I loved the game more," he would say. Then he was drafted into the army. When Banks joined the Chicago Cubs three years later, he had become a new man. He ran the bases hard, he swung the bat with force, he banged long home runs, he dove in the dirt for ground balls. He smiled. He waved. He chattered. He played the game ecstatically. He was the first black man to play baseball for the Chicago Cubs, but his joy transcended color. In the daylight at Wrigley Field, Ernie's joy brought him close to all the shirtless Chicago men who drank beer in the bleachers behind the ivy-covered walls. Ernie's joy brought him close to the men and women who came to the ballgame to get away from the humdrum of daily life. Ernie's joy brought him close to all the fathers and sons in the stands who dreamed of playing big-league ball. They dreamed of playing ball like Ernie Banks.

"I learned how to play the game from Buck O'Neil," Banks would say. Buck said no, Ernie Banks knew how to play, but what he did learn was how to play the game with love. Banks began each baseball game by running up the dugout stairs, taking them two at a time. He then breathed in the humidity, scraped his cleats in the dirt, and shouted what would become his mantra: "It's a beautiful day for a ballgame. Let's play two."

Buck remembered a July game Banks played in Houston. That was 1962 at old Colt Stadium. Buck O'Neil was a coach for the Cubs then, the first black coach in the Major Leagues.

That Houston sun beat down hard on an afternoon double-header. Buck watched Banks run up the dugout steps, two at a time, he breathed in the humidity, he scraped his cleats in the dirt, and he said his bit—Beautiful day, let's play two. Ernie Banks fainted before the second game. That's Houston heat.

Buck was ninety-three years old. People often marveled about his age. Buck never turned down an invitation to speak, and he never said no to a charity, and he often appeared at three and four events a day. And it was amazing: Buck always seemed fresh and alive and young. Only those close to him understood that it was an illusion, that he worked hard to stay young. He took catnaps on the car rides between appearances. He ate two meals a day as he had for seventy-five years. He often showed up for an event, waved to the crowd, spoke for a few minutes, and then excused himself. "Where did Buck go?" people would ask. By the time they had noticed him missing, Buck had already collapsed in his hotel bed.

Something else invigorated him—something harder to describe. It was the thing I found myself chasing all through our road trip. That day in Houston, Buck had signed autographs and told stories and posed for photographs. By the time the ballgame started, he was already exhausted. By the second inning, the sun had beaten him down too. Buck announced that he was ready to go home. Then something small happened. The Houston right fielder, Jason Lane, tossed a baseball into the stands at the end of an inning. The ball landed a few rows down from where we were sitting. Two people reached for the ball. One was a thirty-something man in a sports coat and loosened tie. The other was a boy, probably ten or eleven. The boy wore a Houston Astros jersey with the number 7 on it. Buck always loved baseball numerology. Number 7 was particularly magical—it was Mickey Mantle's number. In Houston, 7 belonged to Craig Biggio, a scrappy, hardworking player. Biggio was a Buck O'Neil kind of player.

The boy and the man both stretched for the ball, but the man was taller and he had the better angle. He caught the ball. He threw his arms up in the air, as if he was signaling a touchdown. He showed the ball to the people around him. He did some variation of the "I got the ball!" dance you see at ballparks. The man was happy. The boy was glum, and he sat down.

"What a jerk," I said.

"What's that?" Buck muttered.

"That guy down there caught the ball and won't give it to a kid sitting right behind him."

Buck looked down and—on cue—the man showed his new baseball to his neighbors. He talked at a feverish pace. Even though we were a few rows back and could not pick up on what he was saying, I had no doubt he was recounting his catch, and I had no doubt that the longer he talked, the more dazzling his catch would become. Everyone likes to believe they're the hero of the story. In this guy's mind, the story was not: "Hey, look at me, I'm the jerk who took this ball away from a kid." No, in his revisionist history, he had to jump up to catch the ball. He had to stand on his chair. He had to catch the ball to save a baby. Maybe he had to dodge snakes and avoid rolling boulders. By the end of the game, I suspected, he would make his catch seem on par with the catch made by Al Gionfriddo, "the Little Italian," who went back to the wall in the 1947 World Series and snagged a Joe DiMaggio smash, spurring the Great DiMaggio to kick the dirt in disgust. The man in the sports coat and loosened tie looked proud as he relived his heroics. One row back, the kid in the number 7 jersey moped while his father mussed his hair.

"What a jerk," I said again.

"Don't be so hard on him," Buck mumbled. "He might have a kid of his own at home."

That stopped me cold. A kid of his own. I had not thought of that. I looked hard at the man, who now wrapped his fingers across the seams of the baseball. He appeared to be showing his friends how to throw a curveball. *A kid of his own.* True, the man did not seem the father type. But it was possible. I tried to imagine this man's kid sleeping at home—a little boy perhaps sleeping on Houston Astros bedsheets. I tried to imagine the boy's thrill the next morning when he woke up, got out of bed, rubbed the sleep from his eyes, and then looked toward his dresser, and . . . what's this? A baseball! White! Glowing!

Did you catch this for me, Dad?

You betcha. It was a one-handed grab! I had to dodge a snake! And later, if you finish your homework, we'll go out and throw that ball around. I'll teach you how to throw a curveball.

Would you, Dad? That would be so great!

I tried, as I would the whole road trip with Buck O'Neil, to see things through his eyes. For five seasons, I would watch Buck look at the bright side. He had every reason to feel cheated by life and time—he had been denied so many things, in and out of baseball, because of what he called "my beautiful tan." Yet his optimism never failed him. Hope never left. He always found good in people.

"Wait a minute," I said to Buck. "If this jerk has a kid, why didn't he bring the kid to the ballgame?"

Buck O'Neil smiled. He was not tired now. He looked young again.

"Maybe," Buck said without hesitation, "his child is sick."

And I realized that no matter how hard I tried, I would never beat Buck O'Neil at this game.

BUCK O'NEIL'S
AMERICA

Buck O'Neil never asked me what this book would be about. This was good because for the longest time I did not know. This book has been many things over time. It was, at first, a story about one fascinating Negro Leagues baseball game played in 1939. The game featured two pitchers and rivals now in the Baseball Hall of Fame— Satchel Paige and Hilton Smith. I researched that game tirelessly, but I could only find two newspaper accounts of it, a total of three hundred eighty-two words. The only player from that game still living was Buck O'Neil.

"I just don't remember much about that game," Buck said. He felt bad, as if he had let me down. We had known each other for a few years by then. I was a newspaper columnist in Kansas City and Buck was Kansas City's favorite son. I wrote about him often. We grew close. It was an easy arrangement. Buck liked telling stories about his baseball days as a player and manager in the Negro Leagues and as a coach and scout in the Major Leagues. I liked to listen. After a while, Buck asked me to lunch. He said that somebody needed to write about what the Negro Leagues were really like. He said, in his deep and lyrical voice, that Hollywood got it all wrong. The

Hollywood filmmakers had stretched truths and diminished legends into caricatures, and in general they made the players look like fools. And Buck said the books about the Negro Leagues mostly read like encyclopedias, and that was no way to get people interested. The books, he said, treated players like cold history, when in fact the players were hot, flawed, brilliant, and alive.

"Somebody needs to write that book—the one that tells what it was really like," he said, and he put his hand on my shoulder. "You'll do it," he said.

Don't get me wrong. It wasn't like Buck knighted me. I suspect he had given this precise speech to other writers. Still, I tried for Buck. I spent dozens of hours reading newspaper clippings from the old black newspapers—the *Chicago Defender*, the *Pittsburgh Courier*, the *Kansas City Call*. I learned about famous men called Satchel, Cool Papa, and the Devil, and also less famous men called Pee Wee, Turkey, and Slow. I talked to former Negro Leaguers, though it often seemed they were dying faster than I could interview them. Still, my book about the baseball game in 1939 was doomed. There were too many empty places and unanswered questions, and I found myself guessing about things I could never know. The next version of this book, then, was a Negro Leagues novel, a made-up story of a white baseball scout sent into the black leagues. That book died for the opposite reason; it was too real to be fiction.

I put the book away for a while. Every time I saw Buck, which was plenty often, he would ask me how I was doing with the book. I told him I wasn't sure. He nodded, and he quickly changed the subject to politics or the latest scandal. He was disappointed. It wasn't only Buck. My wife, Margo, asked often about that book too. I would sometimes mention to her another book idea, one that somehow seemed easier, maybe a book about someone still living or a game I had seen or something like that. She shook her head. "What about the Buck book?" she asked. That's what she called it. The Buck book.

I said: "I don't know how to tell it."

She said: "Maybe you are trying too hard."

When you're a sportswriter, it is easy to grow jaded. You get comfortable seats in air-conditioned boxes, and you usually get a nice parking spot close to the door. You wear plastic-coated credentials around your neck, and you talk to ballplayers every day, and though you may not get to know them—not like sportswriters and athletes in the old days, before money separated them—you do get to know some things. You get to know that they're human, they ache, they curse, they grumble, they preach, they dote on their children, they cheat on their wives, they take amphetamines to perk up for games, they give to charity, they party, they smile at babies, they forget how people who don't make millions of dollars live, they live quiet lives, they take steroids to get stronger, they play heroically, they play halfheartedly, they drink, they lie, and they do small kindnesses when no one is watching. Athletes are really just like everyone else.

That simple fact can disappoint you if you're not careful. Everybody knows that sportswriters get cynical. They turn cold to the games they once loved and the athletes they once idolized. I used to worry about that happening to me sometimes. And one day, while worrying about it, I realized why I was so drawn to Buck O'Neil. He was almost ninety-four years old. He had been denied his whole life. He wanted to play Major League baseball, but because he was black he never did. He wanted to manage in the Major Leagues, but again, he was black and he was never given a chance. He became a baseball scout, and he drove his car all around America in search of talent, and he never got much money or credit. He told stories about the great Negro Leagues players of his time—"We could play!" he would shout—and for many years few wanted to listen or believe.

And yet Buck persisted. He still loved baseball. He loved people. He forgave, but so easily that it hardly seemed like forgiving. All those bad memories faded into nothingness, and in the place of muffled dreams, Buck recalled hot jazz in smoky bars and steaks that melted in his mouth like cotton candy. He remembered Count Basie playing "Jumpin' at the Woodside" on Saturday nights. Sunday mornings he would come downstairs for breakfast in the hotel, and Joe Louis would be eating with Billie Holiday at the next table. After church, Buck's team, the Kansas City Monarchs, would play baseball in the bright sunshine while adoring fans in their Sunday best ate barbecue and cheered. He remembered the sensation of scooping low throws out of the dirt and the way the bat buzzed and vibrated through his arms when a pitch hit too close to his hands. He remembered the feeling of rounding second, heading to third, his heart beating through his chest, his breath caught in his lungs, and the third-base bag looking about as far away as Paris. To find the bitterness in Buck, you would need one of those explorer boats they use to find treasure on the _Titanic_.

"How do you keep from being bitter?" I asked Buck. In fact, I asked him that dozens of times. Everyone asked that same question—cabdrivers, politicians, talk-show hosts, players, executives, schoolteachers, and children. That day, Buck's answer sounded something like a song. I had noticed before how Buck's words often sounded like songs. He said:

> *Where does bitterness take you?*
> *To a broken heart?*
> *To an early grave?*
> *When I die*
> *I want to die from natural causes.*
> *Not from hate*
> *Eating me up from the inside.*

I asked Buck if he would come with me to a basketball game. Buck said no, he was going to make an appearance in a small Kansas town called Nicodemus. Buck made two hundred appearances a year, most of them to help publicize the museum he had helped to create, the Negro Leagues Baseball Museum. Most of his speeches were in big cities. Sometimes, though, a small town held a day in his honor, and Buck never turned down a chance to talk about the Negro Leagues. I asked him if I could come along to Nicodemus. He said: "Be on time."

WHAT WAS YOUR best day in baseball?" Buck asked me on the long car ride to Nicodemus. I would find Buck did not like talking on long rides, but he did not like silence either. So he threw out unexpected questions to keep conversation alive. Who was the best shortstop ever? Why don't more people ride in convertibles? Has anyone ever gotten better at golf by reading golf magazines? Do you think this president will make our country safer? Who was a better hitter, Ted Williams or Barry Bonds? He would ask a question and then sit back, stare out the window until the talk had dissipated, and then he would ask another.

Buck asked about my best day in baseball, and I thought about many things. I had a few flickering memories from my days as a Little League baseball player. I remembered diving stops I had made in a championship game for an underhand-pitch team called Hollywood. I remembered hitting a game-winning double with a wooden bat—I refused to use the aluminum bats. Everybody in the Major Leagues, swung wood. I remembered my one Little League All-Star game appearance, though my lasting memory from that day had nothing to do with the game and revolved instead around teammates teaching me how to cross my legs like a man—right leg bent at a ninety-degree angle, right ankle over left knee. I had been doing it wrong.

I thought then about my best baseball days as a sportswriter and fan. I was in a makeshift ballpark in Charlotte when the most amazing athlete, Bo Jackson, hit his first professional home run. He broke his bat on contact, but the ball soared over the Krispy Kreme doughnut sign in left field. I was in Atlanta for a playoff game when Sid Bream rounded third and scored ahead of a Barry Bonds throw, and the stadium rocked and the cheers were louder than any I had ever heard. I was in Boston when the Red Sox won a World Series game—a lovely woman named Loretta Kowal wore a green Red Sox cap because she was born on St. Patrick's Day. Loretta had cancer. She said she was just thankful to live long enough to see the Red Sox win a World Series game, and she died a few weeks later. I was in Greenville, South Carolina, when a tall Minor League baseball player named Michael

Jordan looked bad. He struck out two or three times. He was hitting about .200 at the time. The scouts, in their unfeeling way, called him a "non-prospect." That meant he would never play in the Major Leagues. So why was he there?

"Have you ever had a dream?" Jordan asked.

I was in New York when Derek Jeter hit a home run just after midnight. It was less than two months after September 11. After Jeter rounded the bases, fifty thousand New Yorkers sang "New York, New York" with a crackling recording of Frank Sinatra. Then they sang it again and again. After six renditions, the public-address announcer asked everyone to leave, but his voice lacked conviction, certainly compared with Sinatra's, and anyway this was New York, where people did not mind public-address announcers. The New Yorkers stayed and cheered and sang about those little-town blues melting away and being a part of it. I was in St. Louis on a warm September night when Mark McGwire hit a low line drive that slipped over the left-field wall. Thousands of camera flashes popped at once—close your eyes in that instant and you could still see the echoes of light flaring in the dark—and McGwire, in childlike excitement after hitting the home run that broke Roger Maris's cherished record, missed first base. McGwire ran back, stomped on the bag, then raced around the bases to the kind of cheers generals hear when they liberate cities. When McGwire reached home, he lifted his son high in the air. Even now, even though I presume that McGwire was chemically enhanced when he broke the record, I feel a little something building in my throat when I think of that night. Fathers and sons and all that.

Instead of all that, I told Buck about my father, Steven, born in the Soviet Union during World War II. He and my mother moved to America three years before I was born. His accent never thinned, and he never quite broke away from Eastern Europe. My dad won the Cleveland Chess Open one year when I was young, and he tried vainly to teach me how to mate a king with pawn and knight. He played soccer semiprofessionally for a while in Poland, and he—again, in vain—tried to pass along the physics of trapping a soccer ball between foot and earth.

Still, beyond the soccer, chess, and accent, there was something distinctively American about my father. He drove us around the neighborhood every year on the Fourth of July so that we could see children running with sparklers and hear firecrackers popping and cracking over the clatter of our 1976 Chevy Nova. On Thanksgiving, we would go around the table and say what we were thankful for in the way of some family on one of those television sitcoms we all watched every night. On Labor Day, we gathered around the television and watched the Jerry Lewis telethon, not because we wanted to see Lola Falana dance but because in my parents' imaginations that's what American families did on Labor Day.

In those Cleveland evenings fluttering with mosquitoes, my father and I played catch. We threw the baseball back and forth in our small and fenced backyard, under the telephone wire that drooped a bit lower to the ground every summer, near a hole to China and an ant-infested picnic table and a grill that always smelled of leaking propane gas. When I was a toddler, my father had bought me a cheap glove. The glove was plastic, and it was packaged in plastic with a plastic Wiffle ball bat and a plastic ball. After he learned more about baseball, he took me to Kmart and spent a few hours of his factory salary to buy me a proper glove. He used the plastic one.

"On fly balls, always take a step back first," he would tell me as he smoked his Kents and threw baseballs high over the telephone wire. "It's easier to come in on the ball than it is to go back."

"When you have two strikes, choke up on the bat," he said. "You have more bat control."

"Before the ball is hit, think about where you want to throw," he said.

And so on. He was a Confucius of baseball proverbs. Get in front of the ball. A walk is as good as a hit. Make the easy play. Expect the bad hop. Step into your throw. Keep the bat head up when you bunt. Don't make the third out at third base. If a pitcher is struggling with his control, take a strike. Catch the ball with two hands. Charge the ball. Meet the ball. Follow the ball. Keep your eyes on the ball. I never knew where a semiprofessional soccer player from Poland picked up all that stuff.

"I'm sure he listened to the radio so he would have advice to give you," Buck said softly and without opening his eyes. "Fathers do that."

Fathers do that. My greatest day was not one day, I told Buck. It probably never happened just so. It was a composite. Lightning bugs sparked. Freshly cut grass. Lisa Gottlieb sat in our tree, her canvas Pumas dangling below a hanging branch. Ancient Mrs. Zepkin watered her already soaked garden next door. In the absurdity of memory, I remember not wearing shoes, which is ridiculous, a Mark Twain touch (as is the idea that Lisa Gottlieb sat in our tree and watched). I stood in the far corner of the yard, near the rusted swing set left behind by former homeowners. Night closed in around us. My father threw pop-ups higher and higher against the darkening sky, and he shouted, "Always take a step back first," and I circled and spun, caught the ball, and fell into the soft, wet grass. He yelled, "Attaboy!" That was my greatest day in baseball. Buck's face showed no emotion at all when I finished. His eyes were still closed.

"What was your best day?" I asked him. I'd heard him tell it a hundred times. I wanted to know if he was awake.

"Easter Sunday, 1943, Memphis, Tennessee," he said immediately. He opened his eyes. "I was first baseman for the Kansas City Monarchs then. We were playing the Memphis Red Sox. First time up, I hit a double. Next time I hit a single. Third time up, I hit the ball over the left-field fence for a home run. Fourth time up, I hit a long fly ball to right field. As I ran to first, I yelled, 'Hit the fence! Hit the fence!' The . . . ball . . . hit . . . the . . . fence. It skipped past the outfielder. I ran around the bases. My third-base coach called me home. I could have had an inside-the-park home run. But I stopped at third. You know why?"

"Cycle," I said.

"I got the cycle. Single, double, triple, home run. That night I was at the hotel relaxing. My friend Dizzy Dismukes comes up to my room and says, 'Buck, there are some people downstairs I want you to meet.' They were teachers from the local school. I walked downstairs and walked right up to one of those teachers. I said, 'My name is Buck O'Neil, what's yours?' That was Ora. And we were married for fifty-one years."

Buck smiled as he always did when the story ended. "That was my greatest day," he said. "Easter Sunday, 1943. I hit for the cycle, and met my Ora."

"Your day puts mine to shame," I said.

"No it doesn't," he said. "Hang on to your day. If you hang on to that day, you'll stay young. Keep it in your heart. And I will keep it in my heart."

"You'll keep my best day in your heart?"

"Of course, of course," he said. "It's already there. You and your father. All the best days are there."

WE WERE IN New York in summertime. It was Friday afternoon—Buck calls that "the crazy time"—and the city was a frantic mess of car horns, jackhammers, and frustration. People just looked angry. You could see they anticipated the line to the tunnel, cars cutting them off, middle fingers, standstill traffic on a bridge as too many other people tried to escape the drone and drama of the city for the calm of New Jersey, Connecticut, or Long Island. Buck had finished the last of his press interviews. He walked into an elevator going down in the Time-Life Building. A sad young woman stood in the corner. She did not look sad by conventional standards. She had not been crying. No, she looked sad in that distinctively New York City way, the haggard sadness of someone who works in an office too big, pays too much for rent, sleeps in too small a bedroom, and spends too much time getting shoved in and out of a subway car. "There's nothing like New York lonely," Buck had told a reporter earlier that day, and that's how she looked: New York lonely. When Buck O'Neil walked into the elevator, she stared at the ground and tried to make herself small.

He said: "I'm Buck O'Neil, what's your name?"

She looked away and pretended not to hear, as if Buck had asked for change or promised salvation. Buck would not be ignored. *"I'm Buck O'Neil, what's your name?"* She glanced up. She looked sadder than before. She spoke softly and said something that sounded like "Swathy."

"Swathy," Buck said, "you are a beautiful young lady."

She met his eyes. The elevator beeped rhythmically as we dropped floor by floor. Buck talked.

"You're a lucky woman to live in this big city," he said. "I would have loved to live in New York when I was your age. I know you must be doing very well for yourself." She stood up straighter and leaned closer to him as he spoke. When the elevator reached the bottom floor, it bucked for a second and Buck stumbled. She reached out to keep him from falling. Then he held out his arms. "Give it up," he said. Swathy jumped into his arms. She hugged him tight for a long time. Then Swathy bounded out of the elevator and into the city's rush hour. In a thirty-floor descent, her sad look had evaporated like a rain puddle in Central Park.

BUCK ALWAYS SAID the two greatest things in this world are baseball and jazz. Music followed us. Charlie Parker blew his saxophone on the radio. Lionel Hampton played the vibes over the airline headphones. "Listen!" Buck would insist. We listened to a Johnny Cash marathon in western Kansas and a street musician blowing "I Want to Hold Your Hand" in Oakland. Buck danced to 50 Cent in Atlanta, he sang with a gospel quartet in Gary, Indiana, and he sang again with a Hebrew-singing folk group in Kansas City. One warm Saturday afternoon, Buck signed autographs and tapped his toes to Billy Joel's "My Life," which played through static over a speaker nearby.

"I like this," he said.

"Billy Joel isn't exactly jazz," I told him.

"It's all jazz," he said.

Everywhere we went Buck made people happy. In Minnesota, a woman touched Buck's hand and cried. "You can't know what this means to me." In San Diego, a television cameraman stopped filming an interview and instead breathlessly told Buck stories about baseball games he had seen as a child. In Washington, the actress Lynda Carter—Wonder Woman herself—hugged Buck and told him baseball stories. She had never met him before, but she said: "I don't know what it is, but there's something wonderful about you."

To which Buck said: "Good black don't crack." He always said that.

And then, sometimes, he improvised. It's all jazz. Once, in the San Diego airport, while a Muzak version of "Hey Ya!" played, two young lovers kissed on the concourse. They locked together like it was V-J Day in Times Square. People rushed by. Some muttered. Some glanced. Buck walked right over to the couple, and he tapped the young man on the shoulder. Nothing. Buck tapped the shoulder again and the couple disengaged. The young man looked bewildered and dimly angry, as if he had been awoken from a dream. He saw an old man, and he gave Buck his fiercest look, and he prepared to be lectured.

"You must really like her," Buck said.

The lovers looked stunned for a minute. They looked at each other. What did he say? What did he mean? They looked back at Buck, who waited for the words to sink in. Then they broke out laughing.

Buck, satisfied, walked to the exit.

"Who is that?" the woman asked. I told her it was Buck O'Neil. She watched him closely as he stepped on the escalator and dropped out of sight. "He's a great man," she said. "I can just tell."

WE WERE IN Washington in autumn. This was close to the end of the trip. We were in a car again, and rain was falling. Buck talked about 1937, when he played baseball in a grass skirt. Buck played for a team called the Zulu Cannibal Giants. Buck almost never talked about the Zulus. He did not talk often about unhappy things like the owners of the Negro Leagues who didn't have the money to pay their bills, or the stale sandwiches the players had to eat when restaurants refused to serve them, or the bumpy bus rides to little towns that did not want them.

This day, though, he remembered the Zulus. That was a novelty team in the Amos 'n' Andy 1930s—it was half baseball team, half minstrel show. The players dressed in grass skirts, painted their faces, and used bats that were supposed to resemble clubs. Before the games, the players danced in a way the white crowd was supposed to accept as indigenous to Africa. Buck joined the team when he was twenty-five and money was tight. Even now, when he talked about the Zulus, he insisted that he did not dance. The team owner, Charlie Henry, did give Buck a vaguely African-sounding name he long ago shoved into the blank recesses of memory. Buck did play first base in a grass skirt. He remembered that. "We would do anything to play ball," Buck said as the car splashed through the Washington streets. While he spoke, Buck looked out the window, through the rain. In the distance, he could see the top of the Capitol Building rise above the trees. This was why he was talking about the Zulus. This memory was about America.

"We had become conditioned to racism," Buck said. "Hatred will steal your heart, man. You don't have any fight left in you. You accept what's around you. That's what this country was like. We thought it would change someday. We just waited for it to change."

Only he did not say it quite like that. He said it like this:

> *Hatred will steal your heart.*
> *Don't have any fight left.*
> *Accept what's around you.*
> *That's what this country was like.*
> *Thought it would change someday.*
> *Waited for it to change.*

By then I knew what this book would be about. It would be about Buck O'Neil. It would not be a biography. He had written an autobiography already, a good one called *I Was Right on Time*. In it, Buck tells some of his wonderful stories about life, his childhood in Florida, his early days spent romping around America looking for a place where a black man could play ball. Then he talks about his baseball life, his days as a player, manager, coach, scout, spokesman, and all the rest.

But there's something he did not put in the book, something he couldn't. I traveled around America with Buck O'Neil, and after spending all that time with him, I understood better what he meant when he said: "Nobody has written a book about what the Negro Leagues were really like." I think he meant: Nobody had written a book that saw the Negro Leagues and baseball and life through his eyes. When I looked at baseball as a middle-aged sportswriter, I could not help but see steroid hearings and high salaries and expensive tickets and mounting arrogance and an Ayn Rand sensibility that tilts the game to the rich and powerful and the Yankees and leaves the poorest teams, like Buck's beloved Kansas City Royals, standing in the cold with their noses pressed against windows. I couldn't help but believe that in some ways baseball and life used to be better.

We were in Oakland, or maybe it was Dallas or Phoenix, and it was winter. A television reporter asked Buck about the way things used to be. Only she did not exactly ask. No, more like: She told him how things used to be. The players cared more. The game had more heart. People had more passion. The reporter was no older than thirty-five, but she seemed quite certain that everything was better once, long ago, and especially baseball.

"The players used to play for the love of the game and not for money—you know, in your day," she told him. Buck had an amused look on his face. There's no guessing how many times, on our road trip, someone told Buck about "his day." I wondered: What did they know about his day? They knew nothing about riding from one dot on the map to the next—one town named for a former president to one named for an old explorer—and playing baseball on dusty infields against furious dreamers on town teams. They were not there when Buck worked for the post office during the winters, and when he stepped outside for his five-minute break, he would smoke a cigarette, close his eyes against the chill, and think of sun and grass and spring training. And yet Buck never stopped them. He gently corrected them.

"Believe me, we could have used more money," Buck told the reporter.

She did not seem to hear him. People rarely do. Instead she began to talk about her wonderful experience as a young baseball fan. She talked about the feeling of those days when her father took her to spring training games in Florida. She remembered every detail: a hint of orange drifted through the air. The uniforms glowed white. Bats cracked, baseballs skittered through the grass, people shouted, and it was magical.

"I remember catching batting-practice home runs," she said. "That was when baseball was still baseball."

"I don't mean to interrupt," Buck O'Neil said, "but baseball is still baseball."

WINTER

THEY CAN'T TAKE THAT
AWAY FROM ME

Snow melted into puddles on the sidewalks outside. The greatest living ballplayer stood in the darkened entryway of the Negro Leagues Baseball Museum on the corner of Eighteenth and Vine in Kansas City. Willie Mays wore a thick, shiny San Francisco Giants baseball coat, though it was warm inside, and his face glistened with sweat. He stared through chicken wire at a carpeted baseball field in the center of the museum. He would not go to the field. The lights were too bright. "My eyes," Willie Mays said. "My eyes can't take the glare."

The greatest living ballplayer studied the field. There were bases and a scoreboard, and ten bronze statues appeared to be playing a game. Mays tried to guess the names of the statues. He suspected that Satchel Paige had to be pitching and Josh Gibson was the catcher. He guessed that Oscar Charleston's statue roamed center field. Mays had heard all his life about Oscar Charleston. Some of the old-timers who had watched Negro Leagues baseball through the Jazz Age and the Depression said Oscar Charleston—"the Hoosier Comet," they called him—was the best to ever play the game of baseball. When Willie Mays was young—years before he played center field in the Polo Grounds and stickball in Harlem, years before he made catches that appeared like optical illusions and hit 660 home runs in the Major Leagues—he played for the Birmingham Black Barons in the Negro Leagues. It was 1948. The old men in the stands watched him close. They argued among themselves. And they decided that Willie Mays could be the next Oscar Charleston.

"How good was Oscar Charleston, Buck?" Mays asked the old man standing next to him.

"He was you before you," Buck O'Neil said.

Mays nodded as if he had heard that before, and he looked again through the chicken wire at the bronze statues of mostly forgotten men who had played baseball in the Negro Leagues. They had played in a time when black men were banned from the Major Leagues. Segregation was an unwritten rule and mostly unspoken. Every so often some group like the American Communist Party or some rogue reporter would inconveniently ask the question: Why are there no black men in the Major Leagues? At first, the answer went that black men were not good enough players. But these black players consistently beat teams made up of Major Leaguers in exhibition games, and the answer changed. Black players, in the revised explanation, were not smart enough to play in the big leagues or dedicated enough or disciplined enough. That held off the revolution until the Brooklyn Dodgers signed Jackie Robinson, and he broke through in 1947. Willie Mays reached the Major Leagues four years later.

Mays scanned the museum baseball field again and guessed that one of the outfielders had to be Cool Papa Bell. When Negro League players got together they always played the

"Cool Papa Bell was so fast . . ." game. The game would start with someone saying something improbable like "Cool Papa Bell was so fast, he once scored from first base on a bunt." The next player would top him. "Cool Papa Bell was so fast, he once hit a line drive over a pitcher's head and got hit with the ball as he slid into second base." And finally someone (usually Buck) would say: "Cool Papa Bell was so fast, he could turn out the lights, get undressed, and be under the covers before the room got dark."

Mays pointed to first base. He guessed that was Buck Leonard. Buck nodded.

"You know what they used to say about Buck Leonard?" Buck asked. "Sneaking a fastball by him was like . . ."

". . . sneaking sunrise past a rooster," Mays said softly, as if repeating a nursery rhyme he had not heard in a long time.

Mays pointed to the statue at third base and said "Ray." That was Ray Dandridge, the one player on the field Mays had played with. That was in 1951, for a Minor League team called the Minneapolis Millers. Mays was a twenty-year-old kid and by then people in the Major Leagues saw his greatness. Mays played for the Minneapolis Millers for only a few weeks. He hit .477 in that time. Mays was bought by the New York Giants. Mays had grown so popular in Minnesota that New York Giants owner Horace Stoneham bought advertisements in the local newspapers apologizing to baseball fans for taking Mays away.

Dandridge was popular in Minneapolis too, but for different reasons. He was an aging legend. He had played for years in the Negro Leagues and in Cuba and Mexico, and he hit the ball well, but people bought their tickets just to watch him field. They called him "Hooks" because he caught baseballs as if they had hooks on them. He played third base with the grace of a dancer—he lunged and leaped and dove for anything hit near him, and he always seemed to throw out the hitter by a half step. Hooks was thirty-eight when he played in Minneapolis, and he badly wanted his chance to play in the Major Leagues. He could still hit, and of course he could still field, but the Giants had no need for an ancient third baseman and they never bought his contract. The next year, Dandridge played baseball in Oakland, but by then the fastball had passed him by. Dandridge went home to Newark and became a bartender.

"Dandridge helped me become the player I became," Mays said, and then his voice took on an uneasy edge. He had just remembered a television show he had seen. He did not know the name of the show, but on it people brought old stuff they had found around the house, and showed it to antiques experts who would give them a brief history and then estimate the value. *Antiques Roadshow,* Buck said. Mays nodded. He said that on the show some guy had brought in an old Minneapolis Millers jersey. He found it at a church bazaar and bought it for a few bucks. The guy discovered the jersey had once belonged to Willie Mays. The *Antiques Roadshow* appraiser estimated the jersey was worth a lot of money.

"My jersey is selling for eighty thousand dollars," Mays said, biting off each word. That amount had special meaning—in 1959, the year after he had hit .347, won the Gold Glove for fielding excellence, and led the league in stolen bases and runs scored, Mays finally got a contract for $80,000. He had worked so hard just to reach that magical number. He had played Major League baseball like no man before him. He led the league in home runs, and he led the league in stolen bases. He also chased down fly balls nobody else could reach and threw out base runners from just about every spot in center field. In those days, though, owners had all the control; this was long before free agency and arbitration and all that. Owners decided how much to pay a player and there wasn't much any player—not even Willie Mays—could do about it. The Giants paid Mays $80,000 in 1959, and they expected him to be happy about it.

Mays turned away from Buck, he turned away from the photographer who kept snapping pictures of him, he turned away from the field. Mays glared at the wall. His teeth clenched, and his fists jammed hard in his coat pockets. A few minutes before, when Mays was lecturing a photographer for taking too many pictures, Buck had whispered: "Careful around Willie, now. He has a lot of sadness and pain in him."

"Why is that, Buck?"

"Hard being everybody's hero, I suppose," he said.

Willie Mays put on a different pair of glasses and turned back to the chicken wire and the field, but he had lost interest in the guessing game. He leaned on something hard and pointed at the statues. "They're all dead," he said, and it was difficult to tell if he meant this as a question or a eulogy. Buck said, "All are dead but the one you're leaning on."

Mays backed away and realized he had been leaning on a statue of Buck O'Neil. This was Buck as manager of the Kansas City Monarchs. The figure leaned in toward the field, his right hand resting on his hip, left arm folded across his knee.

There was a stern look on the statue's face, as if someone on his team had just done something stupid. It's easy to imagine that a shortstop just let a ball go through his legs.

"Looks like you're mad," Mays said.

"I'd get like that sometimes," Buck said.

"Funny, this is exactly how I remember you, Buck."

"Yeah?" Buck said. "And I remember you in center field running like crazy after fly balls. Your hat would fly off. Yeah! Willie Mays running down a ball in center field, nothing in the world like it. That's what I remember."

"That was a long time ago, Buck."

"It was all a long time ago, Willie."

The tour group had made it all the way around the museum, and the people walked onto the field. They touched the statues. The tour group had been put together specifically for Willie Mays, but he had seen the bright lights and waved them on ahead. He watched them through the chicken wire. He talked about how much money baseball players made today—"What is it now, twenty million a year? Thirty million?"—and what? "I don't have to say it, do I?" Mays said, and he did not have to say it—these players could not run with Mays or hit with him or catch with him or throw with him. Some of these players were great, but the mediocrities got paid a lot of money too, more money in a week than Mays had made in a year. Then Mays said he was not bitter about the money, exactly. It was something else.

"We had fun, Willie," Buck said.

"Yeah, we had fun, I guess."

Buck tried to get Mays to talk about some of the games, some of the players, but Willie Mays had the blues, and he did not speak much. Mays said his eyes hurt. "Glaucoma," he explained. He changed glasses again. "I'd like to go around the museum, Buck," he said, an apology, "but I've got to get my eyes right."

Buck nodded. Mays leaned toward the exit. Buck would say in that moment he looked at Willie Mays and remembered watching a game long ago, Buck could not remember where. He saw a ball hit hard, a line drive into the gap between left field and center. He thought: *Well, there's a double.* He looked down at his scorecard to mark it, and then he remembered that Willie Mays was playing center field. So his eyes flipped up and he watched Willie Mays run. Mays was not a graceful runner like Joe DiMaggio or Ken Griffey Jr. or Carlos Beltrán. No, Mays ran with energy and delight, all arms and legs, a child chugging through the sprinklers on a warm summer day. When he was right, Buck said, Mays could outrun his shadow. "There were men faster than Willie Mays," Buck said. "But I never saw one faster with a fly ball in the air."

In this memory, Buck watched Mays run after the ball, run and run, close the gap, his hat flew off, he dove, and then . . . the ball fell just out of the reach of his glove. It skipped away, to the fence, and it was a triple, and on that day Buck knew that winter had come for Willie Mays.

One day it happens
Can't catch up with the fastball
Can't run faster than fly balls
You might lie to yourself for a while
But you can't lie forever
Gotta start a new life
No cheering
No crowds

No teammates patting you on the back
A little piece of you dies.

Mays stood by the door. He looked back on the field one more time. "You know I really don't need to see the museum," he told Buck. "I lived it." Buck said, "Yes you did, Willie. You really lived." And then Willie Mays walked out into the sleet and cold. Someone held an umbrella over his head until he stepped into the car. A few hours later, a close friend of Mays's called and said Willie cried the entire ride back to the hotel. The friend said he rode the elevator up with Mays and walked him to his room, and when they said good-bye and the door closed, Willie Mays was still crying.

NICODEMUS

Buck O'Neil made time. He said that moving is the secret to living. Moving, he said, is the opposite of dying. Buck sank into the leather seat of his dented Cadillac. It was five forty-five in the morning. Drops of rain splattered against the windshield and the beads trickled like teardrops toward the wipers. The windows fogged from the inside. Buck shivered, mostly in disgust. He was going to Nicodemus to talk baseball. His ride was late.

"All right," he griped, "we gotta make time. Let's get this show on the road."

Buck reached out his jittery right hand and turned on the radio. Smooth jazz played. Buck never understood how those two words, *smooth* and *jazz,* ever got together. Celebrity marriage. He snapped off the music. Smooth jazz sounded sacrilegious here. Buck's Cadillac was parked on the corner of Eighteenth and Vine in Kansas City. Through the mist and fog you could see the street sign marking the spot. Jazz fans still took photographs of that sign. Charlie Parker had grown up here, Lester Young, Bennie Moten, Hootie McShann, Big Joe Turner, Mary Lou Williams, magical jazz names. Once, in a bar nearby, late at night, Count Basie and his small orchestra improvised a little Kansas City song for a radio broadcast. "What do you call that?" the radioman asked. They had always called the song "Blue Balls," but Basie knew that would not do on radio. He looked at the clock and saw it was 1 A.M.

"We call it 'One O'Clock Jump,' " Basie said. It became his most enduring classic.

It was like that then. Music poured out of every gin joint on every corner. Kansas City was Boss Tom Pendergast's town then, and it was wide open. The nights were lit by neon, dice rolled, slots spun, ice clicked, whiskey soaked the air, kings and jacks popped in back rooms. Waitresses at Dante's Inferno wore devil costumes, and the Hey Hay Club offered shots of whiskey and marijuana cigarettes for the same price: twenty-five cents. Gangsters, hookers, angels, sharks, suckers, preachers, loners, and ballplayers brushed shoulders. Kansas City jazz was a little bit blues, a little bit gospel, and a whole lot of winging it. The background was a four-beat rhythm, the walking rhythm, straightforward as a Baptist preacher. *BUM-bababa-BUM-bababa-BUM-bababa-BUM.* Sometimes a song lasted an hour. In the wee small hours, after their gigs, the best jazz players gathered near the corner of Eighteenth and Vine, and they dueled with music. The musicians fed off each other's riffs, stole each other's sounds, crashed each other's rhythm, and shouted the blues at each other until past morning. There were no ticket prices for what they called spook breakfasts. Buck said he never again heard anything quite like it.

"I dreamed last night, I was standing on Eighteenth and Vine," Big Joe Turner, the singing bartender, hollered at the Sunset Club, while Pete Johnson banged on piano. Later, in the 1950s, Joe Turner would record *Shake, Rattle & Roll,* which kicked off rock and roll. By then, Eighteenth and Vine was already dead.

Buck O'Neil looked out his Cadillac window—Eighteenth and Vine was still dead. Politicians had tried to revive this corner. They cleaned up the place, tore down some buildings, and lined the sidewalks with streetlamps. They fronted empty office buildings with a new brick made to look like old bricks. This was supposed to make old buildings seem new but look old; it was all very confusing. Nothing moved. Streets looked frozen, like a Hollywood set waiting for a marker to snap and a director to shout "Action!" Rain splashed on slick black streets, and the neon sign of the new Gem Theater reflected off the puddles, casting an artistic blue tint on the scene.

The new Gem Theater was a perfect symbol of what Eighteenth and Vine had become. It had been built to honor the old Gem Theater, where people gathered to watch movies. The only trouble: They did not show movies in the new Gem. It held a few events, a couple of concerts, but mostly it remained empty. It wasn't the same at all. The old Gem had been filled every night. Buck remembered one particular day when he sat in the balcony for a movie at the old Gem. First, though, they showed highlights of a fight between Joe Louis and an actor named Jack Roper. Roper had hoped to stir some interest before the fight by calling Louis an overrated ape. He only stirred Louis's anger. Roper landed one stunning right-hand punch in the fight. This woke up Joe Louis, who promptly pummeled Roper into a trembling heap two minutes and twenty-two seconds after the fight began. Buck could still hear the cheering from the balcony of the old Gem Theater, and it continued even after the movie began. Roper later returned to his acting career. He played in sixty-nine movies, mostly in uncredited roles as men with names like Sledgehammer Carson and Waldo the Wyoming Wildcat. They showed some of those movies at the old Gem.

Buck glanced at his watch again. It was 6 A.M. The driver was now fifteen minutes late. Buck muttered something about being too old to wait on the corner of Eighteenth and Vine. Buck O'Neil was the last of the ballplayers, the last of the 1930s Kansas City Monarchs. The other players with their great baseball nicknames—Satch and Turkey, Streak and Sonny, Too Tall Ted and Bullet Joe—all of them were long dead and buried, uncelebrated, their stories untold, which was why Buck O'Neil sat in a dented Cadillac in an early-morning rain in the first place. He was going out of town to tell their stories again.

"One day I was walking around here with Duke Ellington," Buck began, a story to pass the time.

"We were walking around, and Duke said, 'Buck, let's go listen to some music.' So Duke and I walk into this little club right here on Vine. This whole area, everything, was clubs back then. I don't even remember the name of the club we went into—the place had a million names. What was the name of that place? I'll think of it. We had to walk down a few stairs to get there. And when we sit down, we hear this chubby Kansas City kid blowing on his saxophone. We didn't know what he was playing, you know. He played it fast and wild and all over the place. But you had to listen because it was different, man. You never heard anything like it. You don't hear too many things that are just different."

Buck smiled. He delivered the punch line.

"Charlie Parker," he said. "Charlie Parker. Oh, man. Charlie Parker."

The story ended. There was a point. Buck wanted me to know just how much fun he'd had in those years when he could not play Major League baseball because of the color of his skin.

"People feel sorry for me," he said. "Man, I heard Charlie Parker."

A car pulled up. A man stepped out and hurried over to the window. He said his name was John, and he apologized for being late. He mumbled something about traffic, and then he held an umbrella over the window to keep Buck from getting wet. Buck got out, stepped out from under the umbrella, and walked in the rain. "How do you get to Nicodemus?" Buck asked.

"Well," John said. Then, as if he had rehearsed the line, he said: "You go straight until you get to nowhere and then you turn right."

"All right, then," Buck said. "I've been to nowhere before."

JOHN, THE DRIVER, was a manic radio fiddler. His impatience burned. Even before he had driven us out of the Kansas City metropolitan area, John had already spun through five or six radio stations. Over the sounds of Smokey Robinson and the Miracles, John announced that he did not know what to do with his life. Apparently, he was a life fiddler too.

"Buck," he said, "I still haven't figured out what I want to do."

"Some people it takes longer than others," Buck said.

The drive from Kansas City to Nicodemus took five hours, though with the unvarying Kansas backdrop of wheat, milo, and Casey's General Stores, it could feel like ten days longer than forever. Buck had not traveled western Kansas in fifty years, since his days as a barnstorming ballplayer with the Kansas City Monarchs. Even so, he remembered playing baseball in many of the towns on the highway signs—Council Grove, Abilene, Great Bend, Russell, Hays. He remembered playing town teams on hard dirt infields while a hot Kansas wind swept through them. He said the roads were bumpier then. Now, interstate highway I-70 rode as smooth as chocolate silk pie, and when gliding through western Kansas toward the Colorado border, you could feel like you were not traveling at all, like the car stood still and only the clouds and antiabortion signs outside moved. "It looks exactly the same as it did fifty years ago," Buck said. "But at least you can get there faster."

The scenery repeated itself as if on a loop, and John's frenzied radio fiddling entertained Buck for a while. Billy Joel sang about the Stranger for a moment—that face we all hide away forever—and when that dissolved into static, John changed stations and OutKast asked women to shake it like a Polaroid picture. Johnny Cash walked the line. Fragments of football and basketball games faded in and out—it was that season. Buck nodded his head at a steady pace to the ever-changing music. Near Dorraine, John found a radio station playing *The Orchestra Hour.* A song had just ended, and the disc jockey, or whatever such people are called now, said it was an old one, going all the way back to 1938, Buck O'Neil's first year with the Kansas City Monarchs. "That," the DJ said, "was Ben Bernie."

"Ben Bernie!" Buck shouted as if he had run into an old friend. Ben Bernie was an orchestra leader who had gained a surprising sum of fame by leading a big band and, on occasion, shouting "Yowzah!" into his microphone. It was like that in 1938. You could become famous for shouting "Yowzah!"

Another bandleader, Gray Gordon, earned his fame by play-
ing songs while a grandfather clock tick-tocked in the back-
ground. Ants had to dance to Gray Gordon's tick-tock rhythm.
Another bandleader, Kay Kyser, became even more famous.
He called himself the Ol' Professor and had huge nonsensi-
cal hits with his Kollege of Musical Knowledge, songs like
"Three Little Fishies" with its unforgettable chorus:

> *Boop boop dit-tem dat-tem what-tem Chu!*
> *Boop boop dit-tem dat-tem what-tem Chu!*
> *And they swam and they swam all over the dam.*

None of these men led orchestras on Buck's side of town.
Ben Bernie was strictly for the white crowd, which in Kansas
City meant north of Twenty-seventh Street. For black chil-
dren, a swimming pool marked the border. The shrieking
and splashing happened behind closed gates; black chil-
dren were not allowed in. For adults, newspapers divided its
real estate ads under two headings: "North of Twenty-
seventh" and "South of Twenty-seventh." Kay Kyser, Gray
Gordon, Tommy Dorsey, Benny Goodman, and Ben Bernie
played north, often at the Pla-Mor, Kansas City's entertain-
ment wonderland. The Pla-Mor had its own sandy beach, a
dozen bowling lanes, and the biggest dance floor for six
states around. The Pla-Mor dance floor was cushioned by
forty thousand springs. You could dance all night, you could
dance all night and still have begged for more. The kids
danced in marathons. Every Friday and Saturday night, cou-
ples swung and swayed to "Over the Rainbow" and "All or
Nothing at All" and "Stardust," the last written by Pla-Mor
pianist Hoagy Carmichael. Sometimes at the end of long

nights, couples tossed nickels at the feet of those exhausted lovers still holding each other up in the moonlight. "Yow-zah!" Ben Bernie would say. The Pla-Mor was, according to the billboards, the only place to fall in love. One time Cab Calloway, the black bandleader, came to the Pla-Mor to see a show. A policeman cracked him in the head numerous times. The cop's defense in the ensuing lawsuit was that he had told Calloway at least twice that colored were not welcome at the Pla-Mor.

"Music can't be racist," Buck said as *The Orchestra Hour* weakened into static. "I don't care what. It's like baseball. Baseball is not racist. Were there racist ballplayers? Of course. The mediocre ones. . . . They were worried about their jobs. They knew that when black players started getting into the Major Leagues, they would go, and they were scared.

"But we never had any trouble with the real baseball players. The great players. No, to them it was all about one thing. Can he play? That was it. Can he play?"

John muttered something. "What's that?" Buck asked.

"We all bleed red," John said in a voice so quiet Buck seemed to miss the words again. He saw a road sign for Abilene, the town where Dwight Eisenhower grew up. Buck started singing a song that welled up from somewhere in his memory.

> *Abilene, Abilene.*
> *Prettiest town I ever seen.*

Buck seemed surprised when nobody sang along.
Along the highway, white signs burned through the grain-

elevator monotony of Kansas. "The Cost of Abortion is a human life," read one. "Abortion Kills What God Created." "Choose Life: What a Wonderful Choice." On the radio, a woman asked a doctor if the pulsating in her ear was anything to worry about. The doctor did not seem to think so. Oil wells pumped near Russell, where Bob Dole was raised, and on the radio a deep-voiced announcer asked, "What is the passion of Christ?" He paused for a moment and then said: "You are." John switched stations, this time to a Johnny Cash marathon. We heard "Walk the Line" for the second time. "You didn't expect to find rap out here, did you?" Buck asked.

John flipped again to a show called *Swap Shop*. The first caller said he was in the market for wood, mostly two-by-fours. The host of the show asked, "Well, what are you selling?" The man said he had a chainsaw chain sharpener in excellent condition. This seemed to reassure the host. An older woman called in and said she was selling mums and, improbably, a pair of men's sneakers size 9½. "Never been worn," she said.

"You in the market for a chainsaw chain sharpener?" Buck asked John.

"You never know," John said with a small smile.

Buck nodded. "We all bleed red," he said.

To get to Nicodemus from Kansas City, you don't really turn right at Nowhere. You turn right in Hays, which is not very different. Fewer than forty people lived in Nicodemus on the day Buck arrived. Nicodemus was more living museum than town by then. Nicodemus had been one of the first towns west of the Mississippi founded by freed slaves after the Civil War. The founders intended for it to become the largest black colony in America. The brochures explained that Nicodemus had once been a bustling city of eight hundred people and had two newspapers, three general stores, four churches, a school, a bank, and an ice cream parlor! The brochure punctuated "ice cream parlor" with that exclamation point. The school building survived the Kansas wind, so did one church, but the ice cream parlor and town died. The Nicodemus historians explained this had something to do with the railroad not coming through. The first thing Buck did when he arrived in Nicodemus was hug everybody. There was a pattern to Buck O'Neil's embraces. He walked up to strangers and held his arms out, as if holding an invisible tea set. He said, "Give it up." That was it. I never once saw anyone resist. Sometimes people would hold out a hand and try to shake hands. In this case, Buck stepped up his attack. "You can do better than that," he would say. Veteran Buck O'Neil watchers noticed that Buck hugged the women longer than the men, and the pretty young women longer than the pretty older ones. But everybody got a hug when Buck O'Neil came to town.

"Welcome to Nicodemus," a woman said after her hug. She was Joan, and she worked for the National Park Service. Nicodemus was her town. She looked to be about forty, and she wore the drab brown park service outfit. She had planned the day for Buck to the minute. Joan looked quite worried that something would go wrong. She gave the impression that she often looked worried.

"Are you married?" Buck asked first thing, pointing at a bare ring finger.

"No," she said.

"What's wrong with the white boys around here?" he asked.

Joan blushed and smiled. A couple of local men, Fred and Alex, walked over. They wore suits and sunglasses. "We're the Blues Brothers," Fred said, and together—with the alternating rhythm of a vaudeville act—they explained that they often put on their own Belushi-Aykroyd act at various local events and high school football games. Upon close inspection, Fred's suit was blue and Alex's black. They both wore giant sunglasses over their regular glasses, as if they were going to a 3D movie. Buck acted as if he was meeting Belushi and Aykroyd themselves.

"You guys are great," Buck said.

"We'll be driving you into the game," Fred said.

"All right, then," Buck said.

"It'll be an honor," Alex said.

"All right, then," Buck said.

A warm wind blew and rattled the chain on the flagpole in front of the town hall. Spring felt close. Joan and the Blue/ Black Brothers went over the schedule of events for the day, each interrupting the other. Buck would speak about baseball, then he would have some authentic Nicodemus barbecue—every town in Kansas, no matter how small, has its own barbecue—and then the Blue/Black Brothers would drive him to the baseball field, where kids would put on a re-creation of the first recorded baseball game played by blacks. Buck nodded, but he was not listening. He asked: "Who is the oldest person in town?"

"Well, that would have to be Ms. Switzer," one of the Brothers said. "How old is she now?"

"She's one hundred and one," Joan said.

"Take me to her," Buck said.

Joan's worry returned. This was not on the schedule. She looked around helplessly, and Buck put his hand on her shoulder. "It will be all right," he said. One of the Brothers took Buck to a row of small apartments down the street. Through a screen door, Buck could see Ms. Switzer sitting in a recliner. The room was surrounded by photographs and computer-generated banners—"Happy 98th Birthday!" and "Happy 100th Birthday, Grandma!" and "We love you, Happy 101st." She saw Buck and said, "I hear you're famous." She looked as if she had been expecting him.

"I'm just old," he said. He held out his arms—"Give it up." She made no move toward him. He leaned down and hugged her. She said, "Oh my!"

Buck asked her to remember what Nicodemus was like when she was young. She started to tell a story about her mother and how she had been a wonderful cook, but she stopped midway through. Something was bothering her. She asked Buck to sign her autograph book. He tried to get her to remember again. "Tell me about your children," he said.

"They're good to me," she said. "You cannot forget to sign my book."

"I'll sign it, don't worry," he whispered, and he tried once more to return her to better days. "I'll bet you're a good cook," he said.

"You have to sign my book," she said again. This time it sounded like a desperate plea. She reached down for a large spiral notebook bursting with photographs and loose papers and Christmas cards. "I know just where I want you to sign," she said, and she carefully flipped through to an open page. "Sign it to Ora Switzer."

"What did you say your name was, Ms. Switzer?" Buck asked.

"Ora," she said.

Buck kneeled close to her. He said, very softly: "My wife was named Ora. We were married for fifty-one years." Tears welled in his eyes. He tried to look at her face, but she looked only at the book, and distress covered her face. He signed the book and handed it back to her. "Here you go," he said. "It's a beautiful book."

"I'm a hundred and one and a half years old," Ora Switzer said.

"Is that right?" Buck said. "I'm ninety-three."

"I never would have guessed that. People don't come to see me much anymore. I sit here a lot. My courting days are over."

Buck caught her eyes. "Well," he said, "you know, I'm single."

And with that Ora Switzer's face broke, and she laughed in Buck's arms.

WHEN BUCK O'NEIL walked into the town hall, forty-eight people sitting in metal folding chairs stood and applauded. Buck ignored the microphone—he never used them. He leaned over the lectern and said there was no place in the world he would rather be than Nicodemus. Everybody applauded again. Buck spoke for a few minutes about his days as a baseball player. Buck said he wished each of them could have seen the Negro Leaguers play. "We were something," he said. "We played a different kind of game. Fast and loose. Everybody could run fast. You wouldn't know it to look at me, but I could run back then." He asked for questions. A man in the back shouted, "Buck O'Neil! Buck O'Neil! Do you remember me?"

Buck looked hard at the man as if he recognized the face but could not place the name. Buck had perfected this look through the years. At every speaking engagement everywhere in America, somebody asked Buck, "Do you recognize me?"

"Let me give you a hint," the man said. "In 1935, you and the Kansas City Monarchs came to play baseball down the road in Hill City. It was an all-white team. My name is Denny Switzer, and I was the only black player on that team. I was just a kid then. You beginning to remember now?"

"Yes, how are you doing?" Buck asked.

"Satchel Paige was pitching for the Monarchs in that game. And you said to Satchel, 'Try this kid out. See what he can do.' You remember that, Buck O'Neil? You remember saying that?"

Buck began to say something, but Denny Switzer was not finished with the story.

"So Satchel goes ahead and says, 'All right, here comes a pea at your knee.' And I hit a home run off of Satchel Paige. Hit it straight to center field. And when I got to first base, you said to me, 'Son, take a bow, you just hit a home run off of Satchel Paige.' You remember that, Buck O'Neil?"

Buck said: "Yeah, I remember that. You sure that was you?"

Denny Switzer: "That was me."

Buck: "Let me ask you something. When you hit that home run, Satchel just stared at you as you ran around the bases, didn't he? Just stared at you all the way around, am I right?"

Denny: "You're right. That's just what he did. Stared at me all the way."

Buck: "Denny, it's good to see you again. Real good to see you."

Applause. Tears. Buck signed autographs and posed for photographs ("Take your time," he said when he posed with women). He ate ribs served by the only black woman mayor in Kansas. The Nicodemus Blue/Black Brothers had Buck sit in the backseat of a beat-up car that coughed smoke. They drove to the baseball game with a loudspeaker taped to the top of the car, and B.B. King's voice roared with the static. Buck threw out the first pitch of the game and then sat in a tent that flapped in the wind. He played peekaboo with a little boy. He ate ice cream with one of those small flat wooden spoons that look like tongue depressors. "All right, John," Buck said, "it's time to go home."

THE RADIO PLAYED a show called *Gun Talk* on the ride back from Nicodemus. The topic was the end of the assault weapons ban, which the host called "Independence Day." He said: "We'll talk about how these are not really 'assault weapons' and we'll expose other media lies in a few minutes. First your calls." The first call was from a somewhat hysterical man who worried that his hunting guns that had any range would soon be classified as "sniper rifles" by the liberal media and then taken away by the government. The host reassured him. "They're already doing that," he said. A passing billboard read "Abortion Stops a Beating Heart."

"I know I've told you how I met my wife, Ora," Buck said suddenly. He had. That was Buck's best day. He told the story again about that Easter Sunday in Memphis, Tennessee, when he hit for the cycle and walked up to his Ora and said, "My name is Buck O'Neil, what's yours?"

"My name is Buck O'Neil, what's yours?" he whispered again, as if he were practicing. Ora had died of cancer eight years earlier.

"Her parents did not want her to marry a ballplayer, you know," Buck said, "but I won them over. Her father asked me what my father did. I told him: 'He's in recreation.' And he was too. He ran a pool hall. He did some bootlegging. You know . . . recreation.

"I told him I would take good care of his daughter. But really, you know, she took care of me. I was gone so much of the time because of baseball. That's how it goes in this world. Life doesn't turn out the way you think. You just hold on to each other. That's the trick."

He closed his eyes. Outside, the wind blew and the trees swayed and leaned. I told Buck it was remarkable that he remembered that man, Denny Switzer, who claimed to have hit a home run off of Satchel Paige almost seventy years before. Buck shrugged.

"A lot of men hit home runs off of Satchel," he said. "And there are more every year."

"Well, it's amazing that you remembered him. You could see how much it meant to him."

Buck was almost asleep. He said: "I didn't remember him."

Silence. A few minutes later, Buck whispered: "That guy didn't hit a home run off of Satchel Paige. He's too young. When I played with Satchel, that was the 1940s, that guy couldn't have been more than five years old. He might not even have been born."

I asked Buck why he pretended to remember the man. He said:

> *In our beautiful memory*
> *We were all handsome.*
> *We all could sing.*
> *We all had the heart*
> *Of the prettiest girl in town.*
> *And we all hit .300.*

And then Buck O'Neil dozed off, and I listened to him snore.

SPRING

I LIKE TO RECOGNIZE
THE TUNE

Buck O'Neil loved to sit behind home plate and listen to baseball. "Baseball music," he called it. He heard people stomp their feet and clap their hands. That was the beat. He listened to cranky old men beg the pitcher to throw the ball over the plate and vendors shout "Budweiser!" Late arrivals barked, "I believe this is my seat," and children cried for cotton candy, and the public-address announcer blared, "Will the owner of a red Ford Mustang, license plate number . . ." The melody for Buck, though, was the crack of the bat. For fun sometimes, he would close his eyes, listen to the crack of the bat, and guess what happened.

"That sounded thin, it's a pop-up," he would whisper. It was usually a pop-up.

"Line drive," he would say when the sound resounded fully and deeply.

"That's a home run," he would say. He was almost never wrong about home runs. Buck liked to say that home runs echo. He said when you've been listening to the sound of bats all your life, you can hear the echo.

"The crack of the bat sounds different for different players," he said. Buck said there was a particularly loud and full crack of the bat he heard only three times in his life. The first time Buck heard it, he was a child in Florida. He and other friends stood behind the outfield fence in St. Petersburg, and they waited to catch home run balls they could then sell to tourists. He heard this sound, this particular crack of the bat—in his life he would compare it to dynamite blasting and a gunshot. But neither of those quite captured the sound. Buck remembered standing up straight and asking, "What was that?" None of the other kids knew what he was talking about. Then the sound exploded in his ears again. Buck climbed the ladder they had set up behind the outfield wall and looked out on the field toward home plate. He saw Babe Ruth hitting baseballs.

Twenty years later, Buck was a player for the Kansas City Monarchs in the Negro Leagues. His team was playing the Homestead Grays. The way Buck told it, he was in the clubhouse getting dressed for the game. He heard that same crack of the bat—to Buck the sound seemed to shake the walls. He rushed out into the dugout—he was wearing only his jock—and he climbed to the top step. He saw a thick, muscular black man swinging a bat roughly the size of a fully grown oak tree. The man hit the next pitch, and that unmistakable sound rang again in Buck O'Neil's ears. That was Josh Gibson.

Buck, you didn't really run on the field wearing just your jock.

Yes, I did. Of course I did. Someone yelled at me, "Hey, Buck, put your clothes on."

Buck, come on.

It's a story. It's my story.

Fifty years later, Buck was still scouting—he was almost eighty and he still traveled around America in search of tal-

ented baseball players—and he was in Kansas City, on the
field. He was talking to the players. He heard that sound
once more. He turned to look, and there in the batting cage
Bo Jackson hit the ball. Like dynamite exploding but not
quite. That was almost twenty years ago. Buck had not heard
the sound since.

> *I keep going to the ballpark*
> *To listen.*
> *People don't listen.*
> *I will hear that sound again*
> *Maybe today*
> *Maybe tomorrow*
> *Someday. I know it.*
> *I'm not that old.*

Buck had always been a scout at heart. When he played in
the Negro Leagues, he played well, he hit stinging line drives,
and he defended with grace, but mostly he watched other
players. His teammates called him "Cap," even when he was
young, because he just seemed like the team captain. He knew
other players' strengths and weaknesses better than they did.
He shared such good advice that, after a while, players just
gathered around him. When he could no longer play well, he
started to manage in the Negro Leagues, and when the Ne-
gro Leagues began its bittersweet descent into antiquity, he
became a professional scout. He traveled the American South
in search of black baseball players.

Buck was in Montgomery, Alabama, to see a game of semi-
professional players. This was 1968. The players were men in

their early and mid-twenties. Buck understood within minutes that he was wasting his time. Baseball scouts (not all of them, but many of them) begin to rely on a sixth sense. They go to so many games and they see so many players who are not quite good enough—pitchers whose fastballs are just a bit slow, batters who can't quite catch up with even those fastballs—that after a while they believe that real ballplayers will just stand out for them, like those black-and-white photographs with a bright red rose in the middle. Buck saw no roses on that field that day, and he started to leave before the game even began. Baseball scouts (not all of them, but many of them) have a natural impatience. Buck explains it this way: "If the player ain't here, then he's gotta be somewhere else."

But Buck stayed. He had a hunch that something good would happen. It wasn't a feeling he could explain, but Buck, like many scouts, put faith in his hunches. Late in the game, a young player ran on the field. Buck could not take his eyes off him. The kid could not have been older than eighteen. He was small and light and in Buck's memory he did not do anything special in the game. But to Buck O'Neil, there was something about the way he moved.

Even the best scout's hunches lead nowhere most of the time. But the best scouts cannot let go of their hunches. Nobody around him knew the kid's name. He was a backup, they said, some kid from the neighborhood. Buck asked around and finally found out the kid's name was Oscar Gamble. What a name! Oscar Gamble. Buck believed there was something in names. He found Oscar and said, "I'm Buck O'Neil, a scout from the Chicago Cubs. I want to see you play." Oscar was shy then and befuddled; he gave Buck sketchy directions to a baseball field hidden in the woods outside of town. He offered an imprecise time for a game the next day.

Buck got into his Plymouth Fury the next day and drove into the woods to find Oscar Gamble. He found the dirt road Oscar had described, and he drove for what seemed like many miles along that road. He did not see other cars or street signs or tire tracks or any indication that anyone had driven on the road for many years. This was the late 1960s, and no great time for a black man to be driving along a dirt road in the Deep South. When Buck did see a car coming the other way, he had to pull off to the side—the road was too narrow for two cars. The other driver looked at Buck and shook his head.

After a while, Buck felt certain he was going in circles—he had seen this tree and that bush—and he started to turn back. Then he heard something. The crack of a bat. It was not loud, not the explosion of Babe Ruth or Josh Gibson, but it was unmistakable. There was a baseball game being played.

Buck drove around a bend and came upon a little field. Almost forty years later, he still remembered the lush beauty of that field. The outfield grass glowed so green it looked as if children had colored the blades of grass with crayons. The outfield wall was lined with blue ash trees, sixty or seventy feet tall. Buck smelled fried chicken and beer. Families sat around the field in lawn chairs.

From his car, he watched little Oscar Gamble walk to the plate. Again he felt sure he saw something, that indescribable little something that baseball scouts—after enough years of dirt roads, cheap hotels, and baseball games on dusty fields— are sure they alone can see. He watched Oscar Gamble wait for just his pitch and swing the bat with the force of a man twice his size. He watched the ball soar to the wall. The ball was caught there, but Buck O'Neil knew then that he was right. He told the Chicago Cubs to draft Oscar Gamble.

"Who?" they asked.

They drafted him in the sixteenth round and Buck O'Neil signed Oscar Gamble. Oscar hit exactly two hundred home runs in the Major Leagues. He played for seventeen seasons, many of those for the New York Yankees, and he became wildly popular in large part because he wore the largest Afro in baseball. He was not the best player Buck O'Neil signed. But Buck was always proud that he found Oscar Gamble.

Scouting changed in the years after Buck found Oscar. Scouts broke down players' swings on video. They used computer programs to study players' statistics. They traded information on the Internet. More than anything, the scouts started to use radar guns. At almost any game where a promising young pitcher was throwing, you could see the scouts sitting behind home plate pointing their radar guns toward the field like state troopers on the last day of the month. Buck understood that times change, and when you're trying to predict the future—"Scouts are like fortune-tellers," he said sometimes—you use every tool you have. Still, he did not like radar guns. He understood radar guns could tell a scout how fast a pitcher threw, but they could not tell how much life was on the fastball. And that was the important thing. He said:

> *Young scouts point their guns,*
> *Write down the numbers.*
> *Are they watching?*
> *Really watching.*

I wonder if they're looking for life.
Because that's the secret, man.
Miles per hour,
That don't mean nothing.
Does the fastball have life?
Does it move? Does it dive? Does it rise?
Bothers me. Too many scouts
Not watching for life.
Life passing them by.

A BALLGAME IN HOUSTON

A shoeshine man stood outside St. Pete's Dancing Marlin, a lunch spot in Houston. He watched people walk by. The Houston sun beat down; it was not a day for a shoeshine. He stood around anyway, as if he had a hunch something would happen. Buck crossed the street and walked up to the man. He did not want his shoes shined—Buck's blue shoes sparkled. Buck often said that he was such a good shoeshine boy in Sarasota, they still talked about him.

"How are you doing?" Buck asked as he thrust out his hand. "My name is Buck O'Neil."

The man said his name was Skipper, or maybe it was Skitter. Buck nodded. The two breezed into a conversation about Willie Mays.

"The thing about Willie," Skipper or Skitter said, "was that he was so smooth."

"No," Buck said, "not smooth. That's the wrong word. Joe DiMaggio, yeah, he was smooth. Duke Snider. Yeah. Ken Griffey. Yeah. Those guys were smooth. Willie was something else. Willie was exciting, you know."

"I saw Willie Mays make a catch here at the old ballpark. Colt Stadium. You remember old Colt Stadium?"

"Shoot, remember it? Do you remember the mosquitoes there?"

"Oh, yeah. I remember outfielders used to wear towels under their hats like they were Arab sheikhs so the mosquitoes wouldn't get them."

"Those mosquitoes were so big we used to say that everybody should move in groups because otherwise a mosquito might carry one of us back to the nest."

"I saw Willie Mays make a catch at Colt Stadium, man. The ball was over his head, and he just turned and ran back. He outran the ball. He caught it like it was nothing, too. Real easy."

"Great player, Willie Mays."

"Best I ever saw."

"You going to the game today?" Buck asked.

"Naw. I don't go to baseball games anymore. It's changed."

Buck smiled and put his arm around the man. "It hasn't changed," he said. "We've changed. We got older. You ought to go see a ballgame. You're a baseball fan, man. Do your heart good. Help you get young."

"You're right, Buck. You're right. I'll do it."

"You do it."

"And hey, Buck, you stay on the sunny side of the street."

With that Buck smiled, shook the man's hand, and walked toward the ballpark for a game between the Chicago Cubs and the Houston Astros. He sang a tuneless song:

> *Talking to the shoeshine man*
> *On the way to the ballpark*
> *Hot day in early May*

Like old times,
Like the world hasn't changed much.

M INUTE M AID B ALLPARK in Houston used to be
called Enron Field before the name "Enron" suggested some-
thing dark. Minute Maid meant orange juice, which was bet-
ter. As we walked in, Buck talked about the news of the day.
He was particularly taken with the story of a woman who was
supposed to get married in Georgia, but on her wedding day
she called her fiancé from a 7-Eleven in Albuquerque and said
that she had been kidnapped. Evil suspicions focused on ev-
eryone, including her husband-to-be—the family put up a
$100,000 reward for her return—but it turned out she had
made up the whole thing and had taken a bus across the
country. Buck loved this story. He always loved the mindless
chatter of the latest scandal. He read newspapers front to
back every morning, international news to comics, but he lin-
gered on the sports pages and on the scandals. He had been
that way all his life, going back to his playing days. Then, he
read the newspaper on those long, bumpy bus rides between
games.

Other players hated those bus rides. "I lost half my life on
those creaky old buses," a tall old pitcher named Connie
Johnson said. "Those buses would break down three times a
week. I can still feel the rattling in my bones. Sometimes I
shake in the middle of the night and I think it's from those
old rides." Other Negro Leagues players said the same. Buck
rode the same buses—he was Connie Johnson's teammate
throughout the 1940s—but he did not remember a bus ever
breaking down. He did remember the Monarchs bus driver, a

shady character named Murphy. Nobody knew Murphy's first name. He had a mouthful of gold teeth and a mysterious past nobody seemed too eager to uncover.

The players only knew two things for certain about Murphy. One was that he fell asleep at the wheel just about every night. A player was assigned on each trip to jab Murphy awake when his head nodded. "What, what, I'm awake, quit poking me," Murphy would grumble, and then his head would bob again, and one of the Monarchs would say, "Murphy's done fallen asleep again." And the designated player would poke him once more, starting the entire cycle over.

The second thing: Whenever a police car came into view, Murphy would panic and stomp his foot on the gas. He would drive the bus like he was transporting moonshine through the Appalachian Mountains—he almost tipped over the bus a hundred times. Of course, this was a bus filled with bored, edgy ballplayers, so five or six times a night someone would yell, "Murphy, I just saw a policeman!" And no matter how many times they pulled this trick, Murphy would hit the gas and drive like a crazy man until the coast was clear. "I looked for Murphy's face on a poster every single day when I went to work at the post office," Buck would say.

Between Murphy's naps and mad dashes from the cops, Buck read newspapers. Sometimes he read the black papers—the *Chicago Defender,* the *Kansas City Call,* the *Pittsburgh Courier*—but more often he read the white papers. He scoured the box scores of the Major League players ("Look, Hilton," he would say to his roommate Hilton Smith, "Ted Williams got three more hits yesterday. And you say you could get him out?"). He read about wars and elections, deaths and celebrity

marriages, medical breakthroughs, mergers, murders, mov-
ies. He loved Walter Winchell's gossip column. "How you go-
ing to know what to talk about," he asked his players, "if you
don't know what's going on?"

"Do you think the law should do something to that woman
who pretended to be kidnapped?" Buck asked the man sit-
ting next to him. We were in our seats in Minute Maid Ball-
park before the game.

"What can the law do?" the man asked. "She's crazy."

"I don't know about crazy," Buck said. "But she sure didn't
want to get married."

Buck sat in his seat before the first pitch, and he watched
the players warm up. It was an old scout's habit. Buck watched
everything—warm-ups, batting practice, pitchers throwing
in bullpens, on-field conversations. Buck liked watching play-
ers jog in the outfield and he liked to see which player re-
trieved the most baseballs at the end of batting practice. In
baseball, Buck said, anyone can spot the star. But the best
scouts looked for the future star, the starlet in a drugstore.
And to find those, Buck said, a scout had to look for small
moments of grace. Buck looked around the field and took in
the easy tempo of infielders tossing the ball around just be-
fore the start of the game, and he offered a running com-
mentary.

"Look at that third baseman. He's got a strong arm. The
shortstop has soft hands—the ball sticks to his glove, like
Velcro. . . . The second baseman has bad feet. He looks like he's
stomping grapes. . . . The first baseman just scooped a bad
throw out of the dirt, but it was luck. He didn't show that natu-

ral ease. He stabbed at the ball and it was like, *Look what I found*."

Buck loved watching first basemen—that was his position when he played for the Monarchs. "I used to love when my infielders threw the ball in the dirt," he said. "I knew how to pick it." Buck believed that he could watch any first baseman for only a few seconds and know how he learned the game.

"This guy learned the game from his father," Buck said as he pointed at the first baseman.

"How do you know that?"

"I just know."

The pitcher snapped off a curve, and Buck watched to see how the catcher's hand moved. Did he stab for the ball or did he let it tumble into his glove? Outfielders threw. Buck glanced out there to see if the right fielder had that Major League arm. Did the ball jump out of his hands as if yanked by long string? It was a scout's symphony out there, and even though Buck stopped being a scout some time ago, he could not help but listen to the music.

"You see today, but I see history," he said. "That could be Babe Ruth out there, or Josh Gibson, or Willie Mays, or Cool Papa Bell. I see this game a little bit differently. I see a skinny kid swinging with an upper cut, I might say, 'He looks like Ted Williams.' I see a big man pitching, like this big man on the mound, and to me it might be another Nolan Ryan or Hilton Smith. I see little things that remind me. It's beautiful."

A man tapped Buck on the shoulder. "Well, Buck," the man said, "we might have to put you in there today." It was Drayton McLane, the owner of the Houston Astros.

"Well, Mr. McLane," Buck said, "you do need some hitting."

It was true—the Astros were struggling to score runs.

McLane laughed. "He doesn't miss a trick, does he?" McLane said, as if Buck could not hear him, and Buck turned back to the field and focused on that big starting pitcher. Roger Clemens, perhaps the best pitcher who ever lived. This game matched up two great pitchers, Houston's Clemens and Chicago's Greg Maddux. It was the first time in almost twenty years that two men who had won three hundred Major League games faced each other. They won their games so differently. Maddux had won his games with guile and cunning. He slipped pitches over hidden corners of the plate. He never threw the pitch precisely where a hitter expected it. Buck admired Maddux the way he admired classical music. But Buck had different feelings about Clemens, who won his games with power, intimidation, and brute force. Clemens threw his fastball ninety-five miles per hour and aimed for the chin if he suspected a hitter felt at ease. This was the game Buck had played in the Negro Leagues. Pitchers threw at your head. Buck liked to talk about the time he hit two home runs off an old friend named Spoon Carter. The third time up, he knew that Spoon Carter would throw the ball at his head.

"You gonna throw at your old friend?" he yelled to Carter.

"Buck, you hit the ball over the right-field fence and you hit the ball over the left-field fence. Of course I'm going to throw at you." Spoon's next pitch rushed at his head, and Buck ate dirt.

So Buck loved Roger Clemens the way he loved the blues and nickel cups of coffee and shoeshine men and other things time had passed by. Before the game, in a ceremony, Buck had given Clemens a special pitching award named after Buck's friend Satchel Paige.

"It means the world to me to have my name in any way associated with Satchel Paige," Clemens said.

"That's good," Buck said, "because Satchel would have liked you."

Buck watched Clemens closely. "Do you think Clemens can still throw the fastball by people?" he asked the man sitting next to him. All through the game, Buck started conversations with the people sitting around him.

"Most definitely," the man said.

"He's an old man now," Buck said. "You can't throw those fastballs forever."

"He can still do it," the man said.

"We will see," Buck said. Clemens threw the last of his warm-up pitches, and Buck's legs bounced in anticipation of the first pitch.

"Hey," he asked the man, "how much did you pay for that beer?"

"Seven dollars," the man said.

"Seven dollars," Buck said, and he shook his head. "The game never changes. But the prices do."

CLEMENS CAME TO bat in the second inning, and everyone knew he would bunt. Clemens, for all his force as a pitcher, still could not hit a lick. There was a man on first base, one out—this was what baseball announcers called "an obvious bunting situation." The third baseman stood so close to Clemens they could have shared a milk shake. It was insulting, if you think about it, to have a third baseman stand so close. This made Buck think of Earl Wilson.

Earl Wilson was a pitcher who could hit. He had other

attributes as well—he could throw very hard, for instance. He pitched in the Negro Leagues briefly, and then in 1959 he became the first black player to sign with the Boston Red Sox. Later, he won twenty games in a season with the Detroit Tigers. With that, he became the second African-American pitcher in the American League to win twenty—an odd historical landmark, but one compelling enough to be listed prominently in his obituary. Earl Wilson had died just a couple of days earlier.

The obituary did not make much note of Earl Wilson's biggest contribution to the game. No pitcher, with the exception of Babe Ruth, ever swung the bat harder. Wilson rarely connected with the ball, but when he did it flew a long way—he mashed thirty-five home runs in his career. He bunted just twenty-two times, and each of those was against his will.

"It about drives me crazy that pitchers cannot hit," Buck said. "These guys are supposed to be your best athletes, right? When you're a kid and you are a great athlete, what happens? You become the pitcher. So what happens to these guys? I used to say to Earl Wilson, 'Why did you hit so many homers, Earl?' He said, 'I swung hard, Buck.' That's how you're supposed to play this game. Swing hard!"

Clemens shifted uneasily in the batter's box. He looked uncomfortable, as if he had been pulled from the crowd to be part of a hitting contest. "Swing hard!" Buck shouted. "Come on, Roger! Hit it out of here!" Clemens had never hit a home run in the Major Leagues.

"Forget about the strategy, swing the bat!" Buck yelled. "Come on, Roger, be a hero! Help yourself!"

Maddux pitched. Clemens turned and held his bat as if

he would bunt. The third baseman rushed even closer. Then, abruptly, Clemens pulled his bat back and slashed at the ball. He hit it low and hard. The ball skimmed over the infield grass and rolled past the stunned Cubs third baseman. It was a single. Clemens stood on first base, a sheepish, thrilled look on his face.

"Did he do it? Did he do it?" Buck yelled, and he stood up and clapped. "What you say? That's what I'm talking about, Roger! You go! That's baseball!"

Buck settled back into his seat. He offered a quiet commentary on the game.

"No need to cheer that one, folks," he muttered when the crowd shrieked after a lazy fly ball. "That's an easy out."

And to a hitter who complained to an umpire after striking out against Maddux: "Sit down, young man. You're not the first man to strike out against Greg Maddux. You're not the *best* man to strike out against Greg Maddux. Just sit down."

And to a young pitcher who kept throwing fastballs and kept giving up hits: "Put something on the ball. You're not going to be able to throw the ball by them. This is the Major Leagues."

And to another pitcher who kept walking people: "Throw some strikes, dummy."

And when Houston outfielder Jason Lane sprinted toward the right-field found line and made a diving catch, Buck was out of his seat. He screamed, "Did you see that? Did you see that? What a game! What a great game!"

BASEBALL SCOUTS WATCH players' rumps. They study human anatomy. Scouts can talk for hours, often over pretzels and beer, about a pitcher's arm angle or the way a batter points his toe. And they will try to unravel all of baseball's mysteries by watching the twitches and turns of a batter's gluteus maximus. If a hitter is confident, Buck said, his rump will stay put while he swings the bat. Playing and scouts call that "staying in there." But when a hitter feels unsure, when he's been expecting a fastball only to be dealt a curve, when his timing has been wrecked by a particularly nasty pitch, his butt will go flying one way or another.

A Houston player named Willy Taveras came to the plate. "Man, he looks like Willie Mays," Buck said, and he meant that literally. He thought Taveras's face bore a resemblance to Mays's face. As a hitter, Taveras looked nothing at all like Mays. Taveras's leading attribute was his speed, so his swing was built to chop the ball downward, into the ground, so he could then run hard to first base. Chicago pitcher Michael Wuertz threw his pitch, and Taveras was badly fooled. He chopped with his bat, missed, and almost fell down.

"His heart was willing" was Buck's scouting report, "but his rump was gone."

SOMETHING IS ALWAYS going on at the ballpark. Baseball marketing directors worry about baseball losing touch with the times. They worry that the pace is too slow for the kids raised on video games and the adults who check their e-mails on their phones. So the baseball people constantly invent new things to fill in the quiet spaces between innings and at-bats—trivia contests, dance-offs, beer gardens, children's play areas the size of amusement parks, pizza promotions, and blooper videos that show baseball players colliding into each other. In Houston, there was also "the Great Hummer Race." This was a hot baseball trend—an animated race that would be shown on the video board every night. Every stadium had one. On the Kansas City video board, for instance, animated hot dogs representing those popular condiments ketchup, mustard, and relish raced around the bases. Mustard generally drew the loudest cheers. In New York, animated subway cars raced to Yankee Stadium. In Oakland, animated dots raced. In Milwaukee, real people dressed up like sausages (one bratwurst, one Polish sausage, and one Italian sausage) sprinted from left field to home plate. In Pittsburgh, pirogues raced.

On this night in Houston, animated Hummers plowed through mud and guzzled cartoon gas. The crowd, which included the first President Bush, stood mesmerized and then, when the black Hummer crossed first, the entire stadium erupted in cheers. Houston likes black Hummers.

You might guess all of these baseball sideshows would drive Buck crazy. Truth was, he loved every bit of it. He cheered loudly for the black Hummer. He laughed happily during the bloopers video. In one promotion, a man who looked like he had recently escaped from Folsom Prison was awarded a free "Men's Wearhouse makeover." "I wonder how it will turn out?" Buck said two or three times. It turned out the way you might expect. After the makeover, the man wore a suit, which nicely covered his arm tattoos. He looked as if he had been paroled from Folsom Prison. Buck was impressed. "They really did a number on that guy," he said.

During the seventh-inning stretch, "Deep in the Heart of Texas" played. Buck sang along—of course he sang. There was never a time when I heard people singing and Buck did not join in. In this case, Buck did not know the words, and he could not keep up with the lyrics as they flashed up on the video board. But he tried.

> *The stars at night*
> *are la-la bright*
> *Deep in the heart of Texas*
> *The la-la sky la-la la-la*

Buck's favorite baseball promotion, as always, was the Kiss Cam. This was a ballpark staple all over America. Cameras scanned the crowd and looked for men and women (or boys and girls) sitting next to each other. Music played—always a song like Tom Jones's version of "Kiss" or the Dixie Cups' "Chapel of Love." Once the camera locked in on the couple, one of two things could happen:

1. They kissed.
2. They did not kiss and were booed.

"Mr. O'Neil, can I have your autograph?" a woman asked just as the Kiss Cam was about to begin.

"Yes, of course, dear, in just a moment," Buck said, and then he pointed to the video board. "I want to see this first."

This would be an archetypal Kiss Cam, with all of the usual story lines. First, the camera focused in on a couple of kids, clearly on their first or second date. Unspoken questions rushed through the crowd: Will he have the nerve? Will she? Does she want him to kiss her? Who will make the move? What will happen to them after the cameras turn away? The kids' faces reddened, they giggled, he looked hesitantly into her eyes, she faintly shook her head no, he froze, paralyzed, and the longer they remained on the screen, the more uncomfortable it was to watch. But there's no turning away. "She wants you to kiss her!" Buck shouted.

"How do you know that?"

"I can see it in her eyes."

The boy, being no older than sixteen, saw nothing but clouds in her eyes, and for another two or three excruciating seconds, the boos engulfed them. Her look changed then, her eyes pleaded: *End this.* The tension punctured with a safe peck on the cheek, an uncertain ending. The camera spun away to a middle-aged couple, drunk on beer and attention, and they groped and made out until that too became uncomfortable to watch. The Kiss Cam spotted an old couple and the cheers turned up louder, but apparently not loud enough to jolt him to awareness. He stared off into the distance, and she slapped his shoulder with her purse. He woke from his daze and sprang into action, first shouting "What?" and then leaning over to kiss her. Finally the Kiss Cam found an angry young couple, still not finished with a fight. He drank his beer, she turned away, they would not even look at each other even as the boos swelled louder and louder. She kept looking away. He kept drinking his beer.

"Come on," Buck yelled, "kiss the lady!" He did not.

"Of course, for seven dollars, maybe he should drink the beer," he said. And he turned back to the woman next to him, kissed her gently on the cheek, and signed an autograph.

CRACKER JACK IS utterly intertwined with baseball, of course, mostly because of the song. Every day at every ballpark in every big city and little town across America, people sing "Take Me Out to the Ballgame." In 1908, Jack Norworth, an old song-and-dance man in vaudeville, wrote that song in fifteen minutes while riding the New York subway. In the middle of Jack Norworth's song is perhaps the most effective advertising line in the history of songwriting.

> *Buy me some peanuts and Cracker Jack.*
> *I don't care if I never get back.*

There were problems with Jack Norworth's song. For instance, people who sang "I don't care if I never get back" almost always walked out a few minutes later to beat traffic. The bigger problem involved Cracker Jack—most stadiums did not sell it. Minute Maid Ballpark did not sell it. And yet, somehow, Buck O'Neil ended up with a bag. A big bag. He munched away happily.

"Where did you get that?" a woman a few seats down asked.

"Someone gave it to me," Buck said.

"Are you sure you're supposed to be eating Cracker Jacks?" she asked. There was never a shortage of women who wanted to mother Buck.

"I can eat anything I want," Buck said. "There's nothing wrong with me."

BUCK TOOK MEMORY pills every single day, except for those days when he forgot. He did not know if the pills helped him remember, but when it came to memory he would not take chances. All summer, people asked him: "What's the secret to long life?" He gave many answers, the most common being his standby: "Good black don't crack." But the more anyone was around Buck, the more obvious it was his memory kept him living.

"If Buck ever started to forget, I don't think he'd last long," our constant companion, Negro Leagues Baseball Museum marketing director Bob Kendrick, would often say. "I think that's the only thing that scares him. Dying doesn't scare him. Forgetting does."

The opposite rang true too. Remembering thrilled him. You could see joy flush his face when he recalled a story or a detail he thought lost. An autograph seeker asked Buck if he remembered a ballplayer named Leon Wagner.

"Daddy Wags, of course, of course," Buck said. "He died just last year."

Daddy Wags was a baseball slugger in the 1960s. He was also an actor and a showman. For a while, Leon Wagner owned a clothing store in Los Angeles and its slogan was "Buy your rags at Daddy Wags." Wagner also had a reputation as a man who enjoyed a drink in his time.

"Yes, sir," the autograph seeker said, "I grew up in Los Angeles. He was my favorite player. Did he play in the Negro Leagues?"

"No, no, no, he came later than that," Buck said. "He played at Tuskegee University—I saw him play there many times. Big man. Strong. Hey, listen, I just remembered a story."

That's when Buck's face lit up. The story was about a day game played in Los Angeles. This day game happened a particularly rowdy night for Daddy Wags. He had a hangover, Buck said, that could stop a charging mule. Wags staggered around the dugout throughout the game while his team-mates laughed. Daddy Wags liked the laughter. For some reason, Los Angeles Angels manager Bill Rigney thought it would be a good idea to send Daddy Wags into the game as a pinch hitter. Maybe he wanted to teach Wags a lesson. Maybe he simply did not appreciate Wags's condition. Daddy Wags staggered to the plate. His breath beat him there by three or four steps.

"Daddy," the opposing catcher asked, "how you going to hit with that hangover?"

"Don't worry about that none," Wags said. "One will get between me and my whiskey."

"And," Buck said, "sure enough he hit a ball out of the stadium onto the street nearby. Wags said the toughest part of all was making it around the bases without falling down."

CLEMENS AND MADDUX both pitched well. I asked Buck which pitcher was better. He seemed offended, as if I had asked him to pick his favorite child.

"There is no better, man," he says. "Those are two great pitchers. There is no better. That's just sportswriter talk. They do it different ways. You've got one guy, Clemens, he's all power. You got the other guy, Maddux, and he just knows how to pitch, throw a strike on the corner, throw another one on the corner, yeah, just knows how to pitch. What are you talking about better? They're both great, outstanding, Hall of Fame pitchers."

I rephrased the question. "Buck," I said, "if you were the manager, which of them would you pitch in the seventh game in the World Series?"

"Well," Buck said, "when you put it that way, I'd have to go with Roger."

ONE OF THE Chicago Cubs players, Jerry Hairston Jr., came from a long family line of baseball players. Buck had been around the Hairston family for more than sixty years, going back to Sam Hairston, who was a terrific player in the Negro Leagues. Sam was one of those players who shined in that time just before baseball integrated in 1947. Sam was twenty-seven when Jackie Robinson joined the Dodgers, and he was thirty years old when he signed with the Chicago White Sox. His best days were gone. He managed to make it to the Major Leagues for five at-bats in 1951, and then he went back to the Minor Leagues for good. Sam Hairston, like Buck, became a scout in Chicago.

Sam signed his son Jerry Hairston to play for the White Sox. Jerry Hairston played fourteen seasons in the Major Leagues, almost all of them for the Chicago White Sox. He was a pinch hitter mostly. Nine years after he retired, *his* son, Jerry Hairston Jr., made it to the Major Leagues.

"Fathers and sons," Buck said. "That is what this game is all about. You know what I mean? How many fathers and sons are there in baseball now?"

We thought of a few famous ones: Bobby Bonds and Barry Bonds. Ken Griffey Sr. and Jr. There were other three-generation families too, like the Bells (David Bell; his father, Buddy Bell; and his grandfather, Gus) and the Boones (Aaron and Bret Boone; their father, Bob; and his father, Ray). There were the Alous and Cruzes and Alomars and Fielders and Javiers and Swishers and Wrights and so on.

Buck was right. There were more fathers and sons in base-ball than in other sports. Buck said it has always been that way. He remembered that a son of Josh Gibson played in the Negro Leagues for a while.

"He was named Josh too," Buck said. "Tough thing having that name. He couldn't play a lick either. He must have got-ten his mother's genes."

"Why do you think there are so many fathers and sons in the game?"

"Maybe it's because baseball is a sport you hand down to your kids," Buck said. "Does a father teach his son how to be a running back? No, see, that's instinct. Everybody runs with their own style. Does a father teach his son how to play bas-ketball? Maybe, there are a few fathers and sons in basketball, right? But it's not the same thing.

"In baseball, you play catch with your son. You teach him how to hold a bat, how to swing it, how to get under a pop-up, how to throw to the right base. You teach him how to run the bases. You teach him how to run back on a ball over his head. You teach him how to throw a curveball. You see what I'm saying?"

I nodded. But Buck wanted to be sure.

"In baseball, you pass along wisdom," he said. "Like your father did for you in your backyard."

CHICAGO WON THE game. Buck should have been happy with the result, since he had worked with the Cubs organization for more than thirty years. But we were in Hous-ton, and Buck always wanted the home team to win. He never did like seeing the home fans sad.

On the way out, Buck ran into Ralph Garr, a fine player who stole 172 bases in the Major Leagues. Garr also had a famously shrill and piercing voice. Anyone who heard that voice remembered it. Buck asked Garr about a story he heard—it seemed that Garr was with the White Sox and they were facing Nolan Ryan, who had the Guinness Book record for fastest pitch ever thrown. This was 1978, and Garr led off the game by striking out. He may or may not have seen Nolan Ryan's pitches. Either way, he staggered back to the dugout, looked at his teammates, and said in a voice that could be heard all through the stadium: "Boys, we got *no* shot tonight."

"It's true," Garr told Buck, and his voice again pierced through the noise of the crowd.

Buck was stopped by dozens of people wanting autographs and handshakes and a little baseball talk. Someone asked him how he felt about the news that baseball players were using steroids to bulk up and hit longer home runs.

"People are always looking for an edge," Buck said.

"But I can't understand why someone would risk their health for a sport," a man said.

"You're right," he said. "But think about it. Would you rather be a superstar for twenty years and die at forty-five or be ordinary and live until you're eighty?"

"You're not ordinary," the man said.

"It isn't how long you live," Buck said. "It's how well you live."

IT DON'T MEAN A THING (IF IT AIN'T GOT THAT SWING)

J. C. Hartman reached into his leather satchel and pulled out a manila folder overstuffed with photographs. He placed the folder next to his buttered toast. It was morning. Hartman said: "I remember everything, Buck. I remember every inning of every game I ever played. I remember every hit, I remember every ground ball. I look at these photographs, like this one here. . . ."

A waitress clanked a plate of scrambled eggs on the table. Hartman riffled through his pictures. There were a couple hundred. All were photographs from Hartman's days as a player. He had been a ferocious baseball player for Buck O'Neil's Monarchs in the mid-1950s. Unfortunately, he could not hit. This would seem to be an overwhelming hurdle for a baseball player, but Hartman made up for his weak bat with a crazed gusto that either inspired his teammates or scared them half to death. Hartman never stopped moving or practicing or thinking or talking during games. When the games ended, he still did not stop. Buck had told him, "If you want to make it to the Major Leagues, you have to play harder than the other guys." Hartman played harder. He signed to play in the Minor Leagues, and his zeal for baseball spurred his next manager—a plain-talking Texan named Grady Hatton—to set the North American land speed record for baseball clichés.

Hatton said: "This boy might make the big leagues on sheer determination and hustle. He gives one hundred percent all the time. I'd like to have nine guys on my team like Jay. He doesn't know what it means to quit."

In the hotel restaurant, with the smell of coffee and bacon swirling, Hartman flipped through the photographs. He wanted to show Buck something, but he could not find the right picture.

"These kids, Buck," he went on, "these kids nowadays, they don't know. . . . Hold on, where's that photo? . . . They don't know what we had to go through, Buck. They make more money than God, you know. They make all that money, and they don't know. . . . Let me show you this picture, here. No, this isn't the one. . . . We loved the game, Buck. You know what I'm saying? We loved it. . . . Is this it? No, this one's not it. Hold on. . . . It didn't matter how much money we were making, Buck. Hell, we didn't make any money playing. But we loved it, you know? We loved it, Buck, and I don't see that love now. It's become a business. Here it is, here it is. Look at this picture."

Buck glanced up from his pancakes and looked at a photograph of J. C. Hartman when he played with the Houston Colt .45s. The Major League team was called the Colt .45s in the early 1960s, when cowboys were still the true Texas hero. Soon, though, Americans and Soviets were racing in space, a new kind of hero emerged, and the team changed its name to the Astros. Hartman's face looked thinner in the photograph, but his eyes blazed then exactly the way they blazed now as he talked about baseball. He had made it to the Major Leagues, even though he could not hit. He made it by catching every ground ball, running hard on every play, and trying to win more than anybody else around.

"You know how badly I wanted to play ball, Buck," he said. "You know how much it meant to me. Now, they just want stuff handed to them. They don't want to work for it like we did, Buck. You remember what you told me first time you saw me, Buck? You remember? You said, 'If you don't love this game, you'll never be great.' I remember that like it was yesterday. You gotta love if you want to be great. I was talking to Billy Williams about that."

"How's Billy doing?" Buck asked.

"He's good. He's had some times like all of us, but he's good. Real good. He always asks about you, Buck. He always remembers the time you went to get him in Alabama."

It's a famous story in baseball circles, the story of how Buck O'Neil saved the baseball career of Billy Williams. This was 1959. Buck was a scout for the Chicago Cubs then. He received a phone call from the team's general manager, John Holland. Billy Williams was a top Cubs prospect and Holland had just heard that Williams quit his Minor League team in San Antonio. He went back home to Whistler, Alabama.

"What would you like me to do?" Buck asked.

"Go get him," John Holland said.

Buck drove to Whistler. He knew the way—he knew all the back roads to all the small towns where baseball players lived. He made it to Williams's house in time for dinner. Billy's parents were thrilled to see him. Over dinner, Buck talked about the weather. He talked about the news. He complimented the food and asked for seconds. Everyone at the table kept waiting for Buck to talk some sense into Billy and tell him to go back and play ball. Buck never said a word about baseball. They ate, and after dessert Buck excused himself. He said he had driven a long way and wanted to go back to the hotel to rest. "If it's all right," he said, "I'd like to come back tomorrow."

The next day he ate with the family again, and again he said nothing at all about baseball. In the kitchen, Billy's mother asked Buck when he planned to talk some sense into Billy. Buck interrupted her and said loud enough for everyone to hear: "Billy's a smart young man. He can make his own decisions." Baseball did not come up after that.

Then Buck said: "Billy, I want you to come with me somewhere. Will you take a ride with me?" They got into Buck's Plymouth Fury and rode to a park nearby. Children were playing baseball. When the car pulled up, the kids turned and saw Billy Williams. They rushed over and shouted:

"Billy, it's great to see you."

"When did you get back?"

"Are you hitting a bunch of home runs?"

"I can't wait until I get old enough to play ball myself."

Billy laughed and signed autographs and played catch with the kids and told them stories about playing baseball in the Minor Leagues. When the excitement died down and the children went back to their games, Billy said to Buck: "All right. I'll go back."

That night Buck called John Holland and told him the news. Holland said, "Put Billy on the first plane in the morning." Buck said, "Plane? No, sir. I'm driving him back to Texas myself." They drove back together, and for the rest of his life Williams talked about that ride with Buck and the things they talked about. Billy Williams played ball again. They called him "Sweet Swinging Billy from Whistler." He hit more than four hundred home runs in the Major Leagues. He played every single game for eight straight seasons. When he was finished playing, Sweet Swinging Billy was inducted into the Baseball Hall of Fame. Buck O'Neil was there that day, and Billy Williams winked at him from the stage.

"You know, it's funny, there's one part of that story that always bothered me," J. C. Hartman said. "How did you know all those kids were going to come up to Billy like that? I mean, I've heard you tell that story a thousand times, Buck, and you're a smart man. How did you know for sure those kids would go up to Billy?"

Buck smiled. "Billy was a hero to those kids. I knew they would react that way. Of course"—Buck looked around for a second—"of course, a good scout always checks things out first. I might have gone out there the night before, you know, just to be sure."

"Did you pay those kids, Buck?"

"A good scout always checks things out first."

J. C. HARTMAN SAID: "There's another part of that story you probably don't know, Buck. You know, I was Billy's roommate when he quit baseball. We were the only two black players on the team then. I remember the night he left."

Buck leaned forward and listened. Hartman said they played a game in Victoria, Texas. Williams hit a double in the ninth inning that knocked in the winning run. The game had gone late, and by the time Hartman and Williams finished getting dressed after the game, the one black restaurant in town was closed.

So they went back to the Ambassador Hotel, where the team was staying. They saw teammates eating in the hotel restaurant. Hartman said it was the most familiar scene in the world—ballplayers laughing, drinking, telling stories, a waitress flirting—but on this night it left them cold. Williams and Hartman were black and were not allowed to eat in the hotel restaurant. This was a minor nuisance on most nights because they could eat somewhere else, but now it was late and they were hungry, and they felt humiliated. Hartman found the manager and said, "We need to eat. We're hungry." The manager called the hotel owner. A deal was struck. The manager took Williams and Hartman to the kitchen and set up a table for them.

"You can eat here," he said, and he walked off.

Williams and Hartman sat in the kitchen. They heard their teammates talking about the game and Billy's big hit. Nobody offered them food. A waitress rushed by and did not seem to hear their calls. This was seven years after Ralph Ellison's book *Invisible Man* had been published. They felt invisible.

Billy Williams said: "I can't take this anymore."

Hartman remembered saying: "It won't be like this forever, Billy."

Williams: "I'm not waiting for forever. I'm going home."

Hartman: "Don't do that, Billy. Don't let them beat you."

Williams: "I'm going home. I've got a girl there."

Hartman: "It's going to change, Billy." These were the words Buck O'Neil had instilled in him. Every day, when Hartman played for Buck in the Negro Leagues, he was aware of what it meant to be a black man in America. "It's going to change," Buck used to say.

Billy Williams said: "It's never going to change."

And he left. Hartman did not see Billy in the room that night. J. C. Hartman carried in his mind for forty years a picture of Billy Williams's face that night; his face twisted with rage and frustration. "I couldn't blame Billy," Hartman said. "It was hard. But I knew he'd come back sooner or later. He loved playing. He was just like you and me, Buck. He loved to play."

After breakfast, I asked Buck if he'd ever heard that part of the story before, the part about Billy Williams quitting baseball because he could not get a meal at the Ambassador Hotel in Victoria, Texas. Buck nodded. He said he'd heard it from Billy on the drive back from Whistler. I asked Buck why he never included that when he told the story. He said:

> *Sometimes pain*
> *Is better left behind.*
> *America's a better place now.*
> *Not perfect. But better.*
> *We survived, man.*

BUCK O'NEIL DAY

Buck O'Neil Day began with a 5 A.M. wake-up call in Houston. The car arrived at the hotel at 6 A.M. The flight to Minneapolis was scheduled for seven-thirty. Buck beeped going through the metal detector at the airport, and he had to be patted down by a somewhat embarrassed airline security guard. Buck took off his shoes twice. "I'm sorry, sir," the guard said the second time, though Buck had not complained. "These are the rules."

"It's all right," Buck said. "I understand rules."

A woman asked Buck if he needed a wheelchair, which spurred Buck to bark "No!" a little louder than he intended. There was a winding walk to the gate and McDonald's pancakes for breakfast. The plane was delayed. Buck walked a bit and saw a jewelry store. "I know why they have jewelry stores here," he said. "I'll bet that helps some traveling men get back into their house." He walked back to the gate, and when they called for people who needed a little extra time to get down the jet bridge, Buck was first in line. As Buck's friends said, he never acted old except to get on the plane first.

Buck found his seat and fell asleep the instant he sat down. He woke up next to two middle-aged men who said they had never been on a plane before. Buck was fascinated. Flying had been a part of his life for so long—days like this had become so routine—he simply could not fathom that there were people left in America who had never flown before. The guys explained they had nothing against flying per se. They weren't scared or anything like that. They just had never seen a reason to try it before.

"So what is it that finally got you up in the air?" Buck asked.

"We're going to do some Minnesota fishing," they said.

The plane landed in Minnesota. Two young women—neither of them twenty-five—emerged from behind a train of luggage carts. They could have been sisters. One said they would take Buck to his hotel. They giggled as he talked, and they got lost on the way to the hotel. After a few phone calls and several detours, they found the hotel, which was called The Le Meridien, and it was so fancy it needed the words "The" and "Le" in the name. The rooms looked like something out of a fairly high-budget science-fiction film. High-definition televisions faced the toilets. Lights streamed out of the most surprising places. The chairs in the rooms featured odd numbers of legs—some had three, others five, and some none at all. It took Buck twenty minutes to figure out how to turn on the lights in the bathroom.

"Bob!" Buck shouted to Bob Kendrick. Bob was the Negro Leagues Baseball Museum marketing director. He had helped set up Buck O'Neil Day. Bob had grown up in a small town in Georgia, and he had dreamed of playing in the NBA. When that dream died, he moved into marketing. Bob had big ideas. Like so many others, he fell under the spell of Buck O'Neil. He was hypnotized by Buck's stories about players in the Negro

Leagues. To Bob, those forgotten players stood for something great—they had overcome the worst of America. They played baseball. He started to do some volunteer work for the Negro Leagues Baseball Museum, which at the time wasn't a museum at all. It was a tiny room in an office building in Kansas City. Nobody was allowed to see it, which was appropriate because there wasn't anything to see other than a few yellowed newspaper clippings and a pennant or two. In those days, Buck and some of the other players would pay the monthly rent just to keep the museum alive. After a while, some people approached Bob and said they wanted to expand the museum and turn it into a destination place that people all over America would want to see. It was absurd, but Bob signed on. A couple of years later—spurred by Buck's constant campaigning, Bob's marketing sense, and taxpayer money—a 10,000-square-foot Negro Leagues Baseball Museum opened on the corner of Eighteenth and Vine.

"Bob!" Buck shouted again; it was hard to hear him over the odd New Age music playing in the hotel lobby bar. Bob ate an eighteen-dollar hamburger and fries piled to look like modern art. Mirrored glass surrounded all of us.

"What's that, Buck?" Bob asked.

"What exactly are we doing here?"

"It's your day, Buck. It's Buck O'Neil Day at the ballpark."

"Oh," Buck said. "My day. Yeah."

WHEN WAS THE first time you saw snow?" Buck asked. I told him that, growing up in Cleveland, I was surrounded by snow. In memory, snow first fell in late November, around

Thanksgiving, and we would not see the green of the grass again until St. Patrick's Day. "All that snow seemed normal to you, didn't it?" Buck asked. "Whatever the weather was when you were a child, that's what seems normal to you."

Buck grew up in Sarasota, Florida, and in his memories, a hot sun burned every day. His most brilliant childhood memory, the memory he shared with kids again and again, happened on one blistering summer day when he worked in the celery fields. Buck was a box boy. His job was to take crates out to the celery cutters. It was hard work. The heat overwhelmed him that day, and he shouted out, "Damn, there's gotta be something better than this!" It's the word "damn" that stuck with him. Buck always told the kids, "That might have been the first time I ever used that word. But it was hot that day."

Buck's father was the foreman in the field. He heard his son shout. That evening Buck's father said: "There is something better, son, but you won't find it here. You have to go out and get it." There was no place for an African-American to go to high school in Sarasota. Buck went to Edward Waters College, a black school in Jacksonville, to get his high school diploma. He played baseball and football there. That led him to snow for the first time. Edward Waters College traveled north to play a football game against Morris Brown College in Atlanta. Buck remembered being on the bus, crossing into Georgia, and he closed his eyes and slipped off into a fuzzy sleep. When he opened his eyes, he saw snowflakes drifting in the wind. It took a moment to understand he wasn't dreaming.

"Stop the bus!" he shouted. They all piled out, the whole football team.

"Look at me now," Buck said. He was eating an overpriced

burger in a chic bar in a trendy hotel overpopulated with articles, and he thought of being young and naïve and sticking out his tongue to taste snowflakes in Georgia.

A TWINS OFFICIAL TOOK Buck to the Metrodome in a green Mustang convertible. Buck said: "I was made for this car." The Metrodome was a hideous domed stadium where the Minnesota Twins played baseball. Ballplayers hated the Metrodome. Fans hated the Metrodome. Everybody hated the Metrodome. Baseball was meant to be played outdoors. The light in the Dome always seemed dull. Baseball games seemed to be played behind gray sunglasses. For outfielders, pop-ups dissolved into the roof, and infielders caught their metal spikes in the turf and twisted their ankles. Dan Quisenberry, a great pitcher, summed up everybody's thoughts when he first saw the Metrodome. He said: "I don't know if there's a good use for nuclear weapons, but this might be one."

The Metrodome never inspired affection even on those cold and rainy days in Minneapolis. But on a Minnesota spring day like this—blue and yellow and green, a gentle breeze blowing off the ten thousand lakes—it seemed especially wrong to play baseball here, like children playing in a cramped, smoky basement during summer vacation. Buck caught sight of the Dome and shook his head. "Shame," he said.

Buck walked into the Dome on his day, and a dozen people rushed up for hugs. They gave him the specially made Buck O'Neil baseball card that would be handed out to the fans. They told him how good he looked and how much he meant to them and how honored they felt. "You're lovable," a woman told him.

"I can't help it," Buck said.

They took him to a back room, where he spoke to local high school baseball coaches about life. He went to the lunchroom where cameramen recorded him offering a few words of inspiration for the evening news ("Hank Aaron, Ernie Banks, Willie Mays—these are our genes, man!"). He stepped onto the field to do a radio interview ("Baseball is still the greatest game in the world!"). He went back to the lunchroom and spoke for more cameras ("I always loved playing in Minnesota because here they would treat you like a man"). He went to an empty locker room where an eager young newspaper reporter felt thrilled to meet him.

REPORTER: Do you ever have a bad day?

BUCK: What is that?

REPORTER: Really. Do you ever have a bad day?

BUCK: No. There are no bad days.

REPORTER: But you would have so much reason to be bitter. . . .

BUCK: I stayed at some of the best hotels in the world. They just happened to be black hotels. I ate at some of the best restaurants in the world. They just happened to be black restaurants. In fact, those were better than most of the white restaurants because some of the best cooks in the world at that time were black.

REPORTER: But I guess someone would say to you: How could you not hate?

BUCK: Where does hate get you?

Buck went back to the lunchroom just before the game began so he could eat a few finger sandwiches and celery sticks. The Minnesota Twins general manager, Terry Ryan, walked in with a box of Minnesota Twins golf balls for Buck.

"Hit them like you used to hit baseballs, Buck," Ryan said.

"I wish I could," Buck said.

By now, Buck had been going more or less nonstop for thirteen hours. He did not look tired. He gathered a plate of food and started to eat. A man sat down next to him.

"Buck O'Neil," the man said.

Buck did not look up. He said: "That voice. Tony Oliva, my hero."

FOR AN OLD baseball fan, the name Tony Oliva conjures up the mood and air of the mid-1960s the way hearing Mick Jagger sing "Hey! You! Get off of my cloud" might, or seeing Julie Andrews twirling in the Alps. There's something magical in the name: Oliva. It wasn't that Oliva was the best ballplayer of his time—though there were few better. The name sizzled with possibilities. Tony Oliva burst on the baseball scene in 1964 like no one had before. He won batting championships each of his first two seasons, the first baseball player to do that. Even that does not quite explain things. There was also something exotic about Oliva. He slipped out of Fidel Castro's Cuba by using his brother's passport. His name was not "Tony"—it was Pedro—but he kept his brother's name for political reasons. Once he escaped Cuba, he could not return home to his family. He told reporters in his painstaking and much-practiced English that he called his mother every night.

"I tell my mother I will come home if I am needed," he told the *Sporting News*. "But she tells me to stay here where I have an opportunity to play baseball."

Pitchers did not know how to deal with him. Oliva had a savage swing—the bat often slipped from his hands and spun violently into the outfield or the stands. Nobody threw a bat farther than Tony Oliva. Though he swung viciously, he did not miss the ball much. He hit scalding line drives all over the park. Anybody who watched a young Oliva understood just how hard a baseball could be hit. Oliva in his young days could run and throw too. He was flawless, the perfect player, and he played with the desperation of a condemned man. No one can play the game that hard for very long. Oliva's body broke down. He got the mumps, then chicken pox. He needed seven knee operations. He was a broken ballplayer by the time he was thirty-three. He played on, a decaying designated hitter in his later years, and he still cracked a few hits now and then. But he was not the same. He was not Oliva. There is something melancholy about those athletes who are so good so young.

"You still look young to me," Buck O'Neil said. Oliva's hair had receded and the top of his head shined. His sideburns had grayed and a few dark wrinkles creased under his eyes. But Buck was right. His face still looked young.

"You were as good a hitter as I ever saw, Tony," Buck said.

"I got a few hits," Oliva said.

"Let me ask you something. When you were young, when you were hitting everything, how big did that baseball look coming up there? I mean, it must have looked like a grapefruit."

"It looked as big as the moon, Buck," Tony Oliva said. "As big as the moon."

BUCK AND TONY talked about the split-fingered fastball. They did not believe in it. That's not to say they did not believe it was a good pitch. They did not believe it existed. It was the Loch Ness Monster. It was the Easter Bunny. The split-fingered fastball became popular in the 1970s, mostly because of a relief pitcher named Bruce Sutter. He would throw his splitter by shoving the baseball between his index and middle fingers and then throwing it as hard as he could. The effect made the pitch look like a fastball until the very last instant. And then it would dive down suddenly, violently, like a man walking into a manhole. Hitters would swing right over it. For a long while, nobody could hit Bruce Sutter's pitches. Others started to throw the splitter then. And it became popular.

"The splitter," Oliva said with a smirk on his face. "I guess that's what they call it now."

"It's a nicer name for it," Buck said.

Tony laughed. Buck laughed. Others sitting at the table did not get the joke because they were not old ballplayers. Buck and Tony believed that the split-fingered fastball was, in reality, nothing but an old-fashioned spitball. They even sounded alike. The spitter, unlike the splitter, had been around since the earliest days of baseball. Pitchers back in the nineteenth century had figured out that by spitting on the ball (or cutting the ball, or rubbing the ball with some other foreign substance like Vaseline), they could make it drop at the last instant. The spitball was officially outlawed in 1920, though pitchers often found a way around the rules.

Oliva: "There is no way—I'm telling you, there's *no way*—that you can make a ball drop that much [he held his hands about a foot apart] by putting your fingers on either side. What, do they think I am some kind of fool? There's no way."

Buck: "That's what I been trying to say. The other day, I was sitting behind home plate, and I saw a pitcher throw that big sink. All these young scouts started talking about how that was a great split-fingered fastball. I said, 'Well, that was the wettest split-fingered fastball I ever saw.'"

They each remembered how the pitchers used to get around the rules and throw spitballs back in their day. Buck remembered that on his team, the third baseman was the one responsible for soaking the ball. Oliva remembered facing a pitcher once—even now he was too much of a gentleman to say his name—who loaded up the ball with so much spit that Oliva would get drenched every time he connected with the ball.

"Split-fingered fastball," Oliva said, shaking his head. "People are so gullible now."

TIME FOR BUCK to go down to the field. He hugged Oliva and said, "I love you, Tony." Oliva said, "Buck, you know what I regret? I've forgotten. I wish I could remember like you remember. If I could do it all over again, I would write down every little story. There were so many funny stories, but now that I'm old, I've forgotten."

Buck shook his head.

> *You haven't forgotten.*
> *You just think you have.*
> *Memory is like baseball.*
> *You might oh-for-four today.*
> *But you'll get three hits tomorrow.*
> *Right? Good days and bad days.*
> *You'll remember.*
> *Those stories aren't gone.*
> *They're just behind a few cobwebs.*

Buck winked and walked to the field. Tony Oliva nodded, asked for a pad, and wrote something.

KIDS IN THE stands tossed baseballs to Buck O'Neil for him to sign, and he caught them one-handed. For an instant you could see the athlete he was sixty years before. Buck could always catch the ball. When he was a child at his father's semi-pro baseball games in Sarasota, he would stand in foul ground between innings. People in the stands would throw pennies and nickels at him. He remembered catching every coin. Later, when he played first base for the Monarchs, he loved being in the field. He remembered a game in Chicago where a player lashed a line drive to his left and he dove and caught it. The next batter lashed one to his right, and he dove and caught it again.

"Hit it to someone else!" someone yelled from the depths of the crowd.

While Buck signed baseballs, a woman threw out the first pitch. She had confessed to Buck that she was so nervous she had been practicing for days. He told her not to worry, it would turn out fine. He was wrong. The ball squirted out of her hand sideways and rolled mockingly toward the Minnesota Twins dugout. She buried her face in her hands as people booed. There were not many people in the stands to boo. Twins coach Al Newman walked on the field, stepped behind a microphone stand, and told the crowd about what Buck O'Neil meant to baseball. "He is living history," Newman said. Buck shouted, "Well, I am living!" Buck walked out to the mound to throw out his first pitch. He looked over at the woman and winked.

He stepped on the mound and wound up as if he were going to throw the ball as hard as he could. Then he stopped and jogged forward a few steps.

He wound up again. Stopped. Jogged forward a few steps.

Wound up once more. Stopped. Jogged forward. He gently placed the baseball in the glove of Twins outfielder Jacque Jones, and they hugged as the crowd cheered. The woman said, "I should have done it that way."

"No," a Twins official said, "only one person could pull that off."

Buck walked off with his hands in the air. The Twins manager, Ron Gardenhire, stepped out of the dugout and shouted, "You look good, Buck!"

"I know it," Buck O'Neil said.

BUCK CLIMBED THE stairs and walked around to a table set up for him to sign autographs. The line of people stretched almost all the way around the concourse. He was on his fifteenth hour of Buck O'Neil Day, and he was finally showing signs of wearing down. He signed autographs quietly. He posed for pictures without his usual playfulness. A few people tried to ask him questions. He answered quickly. One woman said, "You can't know what this means to me just to be near you." She did not want an autograph. She just wanted to touch his hands. Buck held her hands and looked like he wanted to say something, but no words came out.

A man from Seattle had tears in his eyes. "You mean everything to me," he said. Buck nodded and signed the next autograph.

"You okay, Buck?" a Twins official asked. He did not answer her. He kept signing autographs mechanically, again and again, a fourteen-second chore kicked off with the rounding of the *B*. The big swooping letters gave him the most trouble. He could run through the next three letters with relative ease, but the last name brought trouble. The *O* in "O'Neil" was the toughest for him—his hand shook and it simply would not let him complete the circle. Sometimes the *O* did come out somewhat rounded, though most of the time it looked like a deflated a beach ball. He struggled with the *N*, labored through the *e* and *i*, and he no longer could put away the *l* with the flourish he had as a younger man.

The line kept growing until finally the Twins cut it off. Unfortunately, they cut it off right in front of a man who wanted to be heard. He shouted that he had brought his wife and child to meet Buck O'Neil—and sure enough, a woman and child stood behind the screaming man, though they did not seem too eager to be connected to him. The man's face blushed crimson. The Twins official said she would try to arrange something, but the man just kept stomping and cursing and shouting that he had brought his family to meet Buck O'Neil, and they would by God meet Buck O'Neil. A few steps away, Buck signed more autographs. He did not seem to hear.

"He's tired," the woman tried to tell the red-faced man.

"I brought my wife and son. . . ."

"He's ninety-four years old."

"And we waited in line. . . ."

"He's been up since five this morning."

"I just can't believe the nerve. . . ."

After a few more minutes of rounding out his *B*'s and *O*'s, Buck signed his last autograph. With that, the woman brought over the red-faced man and his family. Buck looked up and smiled. "It's good to see you," he said weakly. He signed one more autograph.

"Buck," the man said, "I'm your biggest fan."

B U C K   H A D   O N E more chore. He was to sing "Take Me Out to the Ballgame" with his old friend Mudcat Grant. Mudcat had been a good pitcher in Minnesota. He was the first African-American to win twenty games in the American League. He was a hero in the 1964 World Series. Mudcat was also a lounge singer. He used to sing with a group called Mudcat and His Kittens.

Buck sat up in a luxury suite. He looked as if he was going to collapse. His eyes closed. Suddenly he perked up and said, "Hey, look at this pitcher."

He pointed at Andrew Sisco, a very tall pitcher for the Kansas City Royals. Sisco stood six foot ten, making him one of the tallest pitchers in baseball history. But Buck, in his daze, was not pointing because of Sisco's size.

"This guy does a little dance after every pitch," Buck said.

Buck was right. After throwing each pitch, Sisco did a little two-step, a sort of hop-skip. It was not something any of us had noticed, but once Buck pointed it out, it was mesmerizing. Sisco would throw his pitch, stop, and hop. Throw, stop, and hop. Buck said, "One thing about baseball, you never stop seeing stuff."

"Is there anything you miss about the old days of baseball, Buck?"

"No. Not a thing. Baseball is better now than it's ever been. Although . . . Wait, there is one thing I miss. I miss the way everyone used to get dressed up for a ballgame. Yeah. Men used to come to our games right after church, and they looked sharp. The women wore their best dresses and the newest hats, looking pretty, it was something to see. I wish people still dressed up. But you know, times change. The whole world's gone casual."

Buck stepped out of the suite and, in front of a camera, hugged Mudcat Grant, and together they sang "Take Me Out to the Ballgame." Then Buck whispered in Mudcat's ear: "I've gotta go back to the hotel before I faint." Fans stopped him and hugged him on the way out. Buck looked as if he was sleep-walking. The Twins woman pushed him through the crowd. Finally, with the help of a security guard, they broke through and made it outside.

"Wait a minute," Buck said. "I forgot something."

He walked back inside, leaving the Twins woman open-mouthed. "He went back inside," she said. "After all that, he went back inside. What could he have forgotten?" A couple of minutes later, the door opened again. It was Buck. He was eating a vanilla ice cream cone.

"I've got to have an ice cream," he said. "It's my day."

SUMMER

BLUE SKIES

They called him Sonny on account of the weather. Sonny Brown played beautiful baseball in the sunshine. He loved sunny days. Willard Brown, his Christian name, never did play worth a damn on gray days, and he had many gray days in the Negro Leagues. During the week, the Monarchs team bus jolted and bumped over dirt roads and finally pulled into tiny Midwestern towns with names like Liberty and Independence. The bus would jerk to a stop at rock-hard diamonds with scorched grass in the outfield. The Monarchs piled out of the bus. The players already wore their sweat-drenched uniforms. They played a baseball game in a mosquito haze. They faced town teams with factory workers and farmers and coal miners—hard men who long ago may have hoped for something more. Then dust settled, mortgages came due, babies were born, and these men played baseball with sharpened spikes against gifted black men. Busted dreams and moths choked the summer air.

Those games made up the bulk of the Negro Leagues schedules. Teams often played five or six of these sorts of games a week. No one kept records. Sonny Brown never played well on those days. He never could coax out his talents when facing balding white men lashing out against their fates. Sonny kept a *Reader's Digest* in his back pocket, and during games he stood in center field and read. If a ball headed his way, Sonny tossed the magazine to the ground and chased. Sometimes he caught the ball, and the fans laughed and cheered. Sometimes he did not catch the ball, and the fans laughed and booed. It did not matter much to him either way. He walked back to his spot in the outfield, picked up his *Reader's Digest,* and read again.

"Willard," his manager Buck O'Neil shouted from the dugout, "you've gotta be alive out there! You've gotta play the game with joy! Be alive!"

"I'm alive, Cap," Sonny would say. But he wasn't. He was everything but alive on those gray days. He moped and complained and refused to slide on close plays. He walked grudgingly to the batter's box, dead man walking, and he swung at pitches over his head and pitches that skidded in the dirt. He swung at anything and everything, as if he wanted to make the game end. Once, in one of those small towns on one of those gray days, Sonny swung at a pitch that bounced in front of the plate. It bounced. On that day, though, the ball bounced kind of funny and Sonny hit it on the fat part of the bat. The ball sailed out of the park for a home run. The small crowd buzzed—nobody could remember seeing somebody hit a home run on a pitch that bounced. Sonny Brown jogged around the bases with a broad smile on his face, and when he reached the dugout he took off his hat and bowed. Then he winked at his manager and laughed as if the sun had come out from behind the clouds.

"Sonny," Buck O'Neil asked calmly, "why did you swing at a pitch that bounced?"

"The ball was talking to me, Skip," Sonny said. "It was saying, *Swing! Swing!* So I did." Everybody cracked up. Buck too. He tried to look stern. But there was no talking to Sonny Brown on those gray days.

So why did Buck O'Neil love him so much? Why is it that even more than fifty years later, Sonny Brown stood out in Buck's imagination and memory the way Satchel Paige and Josh Gibson and Willie Mays stood out? Easy answer. Sunny days. The biggest Negro League games were always played on Sundays. They played doubleheaders after church. In Kansas City, church normally began in the black neighborhoods at eleven o'clock, but when the Monarchs were in town for a doubleheader, church moved back to ten. Children squirmed in the pews, women fanned themselves with Bibles, and men occasionally coughed: "Speed it up, Reverend." Preachers brought out their baseball sermons on those Sundays— hitters were like Jesus, pitchers like Moses—and when church let out, everyone paraded to the streetcars and the ballgame. The women wore new hats, the men sweated through Sunday wool, and the children ran ahead.

There were no town teams on Sundays. Negro League teams played each other, best against the best, and sportswriters sat in press boxes, statisticians kept the numbers, big crowds attended. On days like that, Sonny ignited. He could do everything under the sun. When the Monarchs played the Homestead Grays, Sonny made a bet with the great Josh Gibson: whoever hit the longest home run would get a steak dinner. In both their memories, Sonny Brown hit the longest home run most of the time. "I can't hit them where you can," Gibson told him.

On those days, Sonny would steal bases standing up. "I've

never seen another man who did not have to slide when he stole a base," Buck would say. Sonny chased down every fly ball in center field. He could do it all. In Puerto Rico, they called him "Ese Hombre"—loosely translated as "that Man"—because there was nobody quite like him. In Kansas City, on Opening Day, the Meyers Taylor Company annually gave out pants to the Monarchs player who hit the season's first home run. Sonny Brown had a closet filled with home-run pants won on Opening Day.

Buck remembered the day in July the St. Louis Browns signed Willard Brown to play in the Major Leagues. That was 1947, the same year Jackie Robinson broke the color barrier. The Browns were a terrible team playing in a town that did not care about them. In a game played one week before the Browns signed Sonny, there were 478 people in the stands. The Browns were desperate, both on the field and off, and they signed Sonny Brown and a Monarchs teammate, Hank Thompson—the first black teammates in the Major Leagues. Browns management hoped Sonny Brown might hit some home runs and, even more, bring in a few black fans. The Browns publicized him as a twenty-six-year-old slugger. Sonny was thirty-two.

"Naturally we believe these colored boys will help us at the gate, especially in our home city of St. Louis," Browns general manager Bill DeWitt told reporters. "Yet we think of them only secondarily as a gate attraction. We are not hiring these men because they are Negroes, but because we hope they can put more power in a club that has been last in the American League club batting most of the season."

"You know one out of eight people in St. Louis is Negro," Bill's brother Charlie added helpfully.

Jackie Robinson played for the Dodgers. Larry Doby played for the Cleveland Indians. Those were the only black men in Major League baseball when Sonny Brown and Hank Thompson signed to play with the St. Louis Browns, the southernmost team in the Major Leagues.

"Good luck to you," Buck said.

"Won't need no luck, Cap," Sonny Brown said. "There ain't nothing but sunny days in the Major Leagues."

He was wrong. The Browns had no sunny days. They had a flying circus. Two days after Brown and Thompson signed, an outfielder from Alabama named Paul Lehner did not show up for the game. He announced that it was not a protest over the signing of two black players. He just had to see a doctor about his leg. Nobody believed him, in large part because it wasn't true. After he was fined, Lehner skipped the next game.

What a team. The Browns had a twenty-two-year-old catcher named Les Moss who hit .157 for the season, one of the worst hitting performances in the history of modern baseball. They had a former child actor named Johnny Berardino playing second base—that season he hit into the rarest kind of triple play where all three runners were caught in rundowns. Berardino many years later became Dr. Steve Hardy on *General Hospital*.

They had a pitcher named Fred Sanford who became known as "the $100,000 Lemon," because that's how much it cost to sign him. He did earn a few seconds of fame that season by becoming the first pitcher in baseball history to turn a foul ball into a triple. He was pitching when Boston's Jake Jones hit a soft roller that rolled foul down the third-base line. Sanford wanted to make sure the ball stayed foul, so he threw his glove at the ball. The glove hit the ball, and according to the rules

of the time, that earned Jake Jones an automatic triple, one
of five he would hit in his Major League career.

The Browns had another pitcher called "Old Folks" Kinder
who drew that nickname because he did not make it into the
Major Leagues until he was thirty-one years old. Things always
seemed to happen to Kinder. Earlier that year, for instance,
Kinder was pitching and a seagull dropped a three-pound
smelt that just missed his head. Kinder, unruffled and appar-
ently used to flying smelt, won the game. It was one of the few
times the Browns would win that year.

And so on. Sonny Brown would say late into his life that he
had walked into a freak show. In Brooklyn, Jackie Robinson
had the support of most of his teammates, a relatively liberal
community, and a brilliant baseball team. The Dodgers' be-
loved Kentucky shortstop, Pee Wee Reese, had made a public
display of putting his arm around Robinson for everyone to
see. Robinson still had to endure endless taunting from
opposing players and fans, and he received countless death
threats. He had friends at least. Sonny Brown's teammates
openly despised him. His manager barely spoke to him. And
the one-out-of-eight blacks who lived in St. Louis did not
come to the ballpark.

Also, there was this: Sonny did not have the heart of a pio-
neer. When Sonny got blue, he played with half his heart. It was
just like those days in little towns. Sonny refused to run hard,
he swung at terrible pitches, he griped to reporters that the
Monarchs were a better team than the Browns. His time in the
Majors could not last. Yes, sometimes the sun peeked out, like
the time in Yankee Stadium when Sonny rapped out four hits.
Most of the time, he seemed lost. He hit .179 in his one month
in the big leagues. Hank Thompson hit better, but by the end

of August, the Browns management still had small crowds and their fill of racial equality. They sent both men back to the Monarchs. Hank Thompson was still young. He would return to the Major Leagues and become a good Major League player. Willard Brown would never get another chance.

Before he was sent back to Kansas City, Sonny did make a little baseball history. That's a story Buck O'Neil told kids in classrooms across America. He said the story was as important, in its own way, as the story of Jackie Robinson turning the other cheek. Sonny did not start either game against Detroit on a doubleheader Sunday, August 13. The Browns had already given up on him. Late in the second game, though, he was sent in to pinch-hit against pitcher Hal Newhouser, who would later be inducted into the Hall of Fame. Sonny was sent in so unexpectedly, he did not have a bat to use. He borrowed the bat of Jeff Heath, the team's best slugger.

Newhouser pitched, and Brown crushed a long drive to straightaway center field. The center-field fence at Sportsman Park was 428 feet away, and Brown's drive smashed right off the number 4. Brown ran full speed, he ran angrily, and by the time the ball made it back into the infield, he had scored, an inside-the-park home run. It was the first home run a black man hit in the American League.

When Sonny walked back to the dugout, nobody shook his hand. Nobody said congratulations. Nobody even looked his way. He sat down and then, Buck O'Neil said, Sonny Brown watched Jeff Heath take his bat, the one Sonny had used to hit the home run. Heath looked at it for a split second and then smashed it against the wall. Sonny Brown talked with Buck about that moment a lot—the moment when he watched pure hatred crashing against a dugout wall.

"What is the lesson of that story?" Buck would ask the children in schools. "He hit a home run and the man broke his bat. What is the lesson of Willard Brown?"

The kids shrugged and had curious looks on their faces.

"The lesson, children," Buck said, "is that it wasn't easy."

NEW YORK, NEW YORK

The black Cutlass crossed the Triborough Bridge just as the sun came up over the skyline. Buck O'Neil sang a tuneless little song.

> *New York, New York,*
> *A town so nice*
> *They named it twice,*
> *Named it twice.*
> *New York, New York.*

"What kind of show am I doing again?" he asked.

"I don't know Buck," Bob Kendrick said. "I just know it's early."

"You gotta get up early in New York City, Bob," Buck said. Buck had come to the big city to talk baseball and spread the word about the Negro Leagues—he loved New York. Buck asked the driver where the Twin Towers had been, and the driver pointed to an empty hole in the sky.

"Where were you when it happened?" Buck asked.

"In the car," the driver said. He said that he heard about the plane crashing into the second tower and tried to call his family, but the cell phone service was jammed. He raced home, and the city was crazy that day. When he finally made it home, he found his family huddled around a phone, and they were crying. "Nine/eleven broke us inside, didn't it?" Buck asked, and he put his arm on the driver's shoulder.

The car pulled up to a door blocked by construction. It was not yet seven in the morning but jackhammers clattered. Buck walked through a tunnel of scaffolding and stepped through a door marked DO NOT ENTER. He came upon a desk where two middle-aged security guards sat. One said, "You weren't supposed to come in that way."

"Well, I'm not from around here. I'm just a country boy from Kansas City."

"You're Buck O'Neil," the other guard said.

"Buck O'Neil. It's such an honor to meet you," said the first. "This is amazing. Buck O'Neil right here. What are you doing here?"

"I'm supposed to do a radio show. Something called *Star* and something or other?"

The security guards' faces changed. They had looked so happy to see him. Now they glanced at each other. Suddenly they both looked ashen.

"Star and Buc Wild?"

"Yeah," Bob Kendrick said, "that's it. That's the show. *Star and Buc Wild*. That's in this building, right?"

"Please don't do that show, Mr. O'Neil. You are a gentleman. Please don't do that. It's the wrong show for you."

"Is that right?" Buck asked.

"They speak ignorance on that show. What do you need to do that kind of show for? You are a great man. There are a hundred radio shows in New York, good shows, where you can talk about good things. Please don't do that show. They speak ignorance."

"Ignorance, eh?" Buck winked. "Well, we'll see if we can talk some common sense with those guys, eh?"

He walked to the elevator. The security guards shook their heads and looked at him sadly, as if they were watching their father go off to war.

THE GREENROOM FOR the *Star & Buc Wild* show was stocked with vodka and tequila and there was one piece of artwork on the wall. It was a photograph of a man who looked as if he had not slept in twenty-three days. He was squeezing a bottle and lotion squirted out the top. The cutline: "Got lotion?" Bob Kendrick started to look nervous. The interview had been set up by a New York public relations firm, and they had not given Bob any advance warning that this interview would be different from the thousand other radio interviews Buck did. Buck had grown so used to the routine: a host would say what an honor it was to meet Buck, to have him on their show, and then they would gently ask him about his life, his memories of the Negro Leagues, and if it was going well they might ask him to tell his famous Nancy Story. It was an easy formula. Something about that photograph and what the security guards had said made Bob think this would be something new.

"I've got a bad feeling about this," Bob said.

"Don't worry about it, Bob," Buck said. "This is New York. You've gotta be tough here."

. . .

Buck was taken into the studio and introduced around. A woman with headphones introduced herself as "White Trash."

"I'm sorry," Buck said, "what's your name again, young lady?"

"White Trash," she said.

"Oh," Buck said. "Well, it's good to meet you."

He shook hands with the man called "Buc Wild." Star introduced himself from behind a table. Buck asked if "Star" was his first or last name, and Star said it was neither, it was just "Star" because he was the star of the show. "Well, that makes sense," Buck said. Bob looked as if he might have a heart attack. White Trash asked if anyone wanted a cup of coffee, and Buck looked for a moment like he might but he did not know how to address White Trash. The small talk ended. The red light went on.

"We're going to take a break from the hating for a while because we have a guest, Mr. Buck O'Neil. . . ."

"It's a pleasure being here," Buck said.

"You've seen a lot of hatred in this country," Star said.

"I've seen the love too," Buck replied.

Star grimaced. And the show began.

Buck did not know this at the time, but Star had been thrown off the radio for a short while a few months before the interview. He'd had an on-air exchange with an Indian woman at a call center.

Star said: "You're a filthy rat-eater. I'm calling about my six-year-old American white girl. How dare you outsource my call. Get off the line, bitch."

He was reinstated. Star's show got good ratings.

S T A R   S A I D   I N   a calm, smooth radio voice: "Jackie Robinson was a sellout, am I right?"

"No, sir," Buck said, "you are not right. Where are you getting that from?"

"He turned on his own people and went to play for the white leagues."

"No, sir. Jackie Robinson was a hero. He was an American hero."

"Look, I'm only forty-one years old. So I wasn't there like you were, sir. I'm trying to be respectful here. But he was a sellout."

"You need to stop saying that. Jackie Robinson was an American hero."

"Well, Satchel Paige and Josh Gibson didn't feel that way, am I right?"

"No, you are not right."

The conversation went on like that for a while. Jackie Robinson was a sellout, no, he was not, yes, he was. After a while, the craziest arguments can begin to sound sane, as fact and fiction blur, and the longer this went on, the more frantic Buck O'Neil became. His voice lifted an octave. Star said Jackie Robinson was a sellout, an Uncle Tom, he betrayed his race.

"No, no, no, listen to me," Buck shouted. "When Jackie Robinson went to the Major Leagues, that was the beginning of the modern-day civil rights movement. That was before Sister Rosa Parks said, 'I don't feel like going to the back of the damn bus today.'" You could tell Buck was getting into the spirit of the fight—he used the word "damn."

"That was before *Brown v. Board of Education.* Martin Luther King was a sophomore at Morehouse at the time. Jackie Robinson went to the Major Leagues, and that's what started the ball rolling. That was the start, man. Are you listening?"

Star looked up. "Some people see it like that," he said softly. "I respect your hustle."

After another moment, Star went to a commercial break. After the red light faded, Star turned to Buck, who was sweating, though the room was not hot. "This is good," Star said. He turned away from Buck and began reading something—maybe it was the latest news or an advertisement he needed to read. Buck looked at Star for a long time.

I'm sorry, the woman called White Trash mouthed, but Buck O'Neil did not see her.

WHEN THE RED lit up again, the conversation seemed to be more peaceful. Star said again he wanted to respect his elders. He repeated that he was only forty-one. He seemed to be winding down. And then he said the Negro Leagues was made up of clowns, like in the Richard Pryor movie *The Bingo Long Traveling All-Stars & Motor Kings.*

"It was nothing like that movie!" Buck shouted. "Bingo Long is not the real story of the Negro Leagues." Star had struck a nerve again. Buck had spent much of his life trying to convince people that the Negro Leagues were real—real players, real games, real joy, real pain. And that movie, a comedy which featured black players shuffling and dancing through towns and often acting like fools, infuriated him. Star tasted blood.

"A lot of people say it was like the movie," he said.

"I was there," Buck said. "It was nothing like it."

"Other people would disagree."

"You're negative," Buck said. "That's why you're here."

"Go ahead and say it," Star said.

"Say what?"

"Go ahead and say it. I'm an uppity nigger."

"Don't use that term. We had to hear white men use that ugly term on us and now we use it on ourselves."

"I'm one of those rambunctious niggers."

"You and I are very different kinds of black men."

Star kept hammering. He brought up a rumor that some white kids were making money on Negro Leagues merchandise. "Don't make this a black-white thing," Buck pleaded. "It's out there for anybody."

Then Star called Jackie Robinson a true sellout again.

"If it hadn't been for Jackie Robinson," Buck said, "so many other things would not have happened."

"He didn't do us any favors," Star said.

Like that, like a hailstorm easing into a summer rain, Star thanked Buck for being on the show. And then the interview was over. The red light went out. Star went back to his reading while others in the room walked over to Buck O'Neil. Apologies danced in their eyes, but nobody wanted to face what had happened. Buck could not see or hear them. He looked right at Star, his tormentor.

"You know something?" he said roughly to Star as everyone

in the room hushed. Star looked up and met Buck O'Neil's eyes.

Buck smiled. "You are my kind of brother," he said.

Everyone cracked up, even Star. In the laughter, Buck O'Neil made his escape.

T H E   S E C U R I T Y   G U A R D S had heard it all on their radio, of course. They would not say, "I told you so." They just nodded when Buck walked out of the elevator. Buck said, "Well, we are awake now."

"The guy's hero is Howard Stern," one of the guards said. "What more do you need to say? I mean, beware of false gods, right?"

"He was a little wild," Buck barely whispered. He tried to shrug it off, but Buck's equilibrium had been shaken, no question about it. He was still sweating profusely, and he did not talk while he signed autographs for the guards. Bob paced angrily. Buck had been ambushed. "Buck," Bob began, and he started to say he would get to the bottom of it, that he was sorry, but Buck shook him off.

"It's good for you, Bob," Buck said. "It's New York, you gotta be tough." He was still shaking, though. While Bob was angry with Star and the public relations people who had set up the interview and a city where a gentle ninety-three-year-old man could be attacked on a radio show, Buck was angry with himself. "I got too comfortable," Buck said. Comfortable was too close to dying.

Two women walked into the building. Buck watched them go toward the elevator for a minute and then he called after them.

"Excuse me," Buck said, "my name is Buck O'Neil. I was hoping I might get a hug."

"A hug?" the first woman asked. "But I don't even know you."

Still, she saw something in those eyes. They always saw something in those eyes. The woman turned to her friend and said, "Oh, what the hell." She threw her arms around Buck O'Neil and he hugged her tight, and after that Buck seemed to have his balance back.

THE NEXT STOP was a television show called *Cold Pizza*. Buck was scheduled to go on after Mr. T, a onetime bodyguard and actor who had become famous mostly for saying "I pity the fool." It was like that in the 1980s. Mr. T still had a Mohawk haircut, and he wore heavy gold chains, and to clinch his interview, he said, "I pity the fool who doesn't watch *Cold Pizza*." Then Buck went on. This interview was a bit softer than the *Star & Buc Wild* show. "You've accomplished so much in your life," the interviewer began. Buck talked about the Negro Leagues. On his way out he hugged every woman in the main lobby. A young man who worked for the station caught his eye while three women were draped over him. "You keep living, young man," Buck said. "You'll get there."

After that, he headed to a radio interview at Rockefeller Center. He told the host there baseball was still the best game in the world. He then remembered playing a game at Yankee Stadium against the New York Black Yankees. He said he will never forget that day because after the game, he rode back to the hotel in a Rolls-Royce belonging to the actor and dancer Bojangles Robinson. Bojangles owned the Black Yankees, and sometimes before games he would race his players to first base. Bojangles would run backward.

"That was the only time I ever sat in a Rolls-Royce," Buck said.

The next interview was in the *Sports Illustrated for Kids* office. He told the editors about his first trip to New York for a baseball game. He did not remember the game itself, but he remembered that afterward he traded in two new suits he had bought in Florida for one zoot suit. He got a chain he would swing around when walking down the streets in Harlem. "I was the hottest thing in New York City," he said.

"What advice would you give to children?" they asked him.

"I'd tell them, 'You can be anything you want to be,'" he said.

Buck then did an interview with WHCR, the Voice of Harlem.

"Why is baseball the national pastime?" the interviewer asked him.

"Because it's for everyone. Baseball is for the pastor. And baseball is for the pimp."

He talked with a reporter from *Source* magazine. She was an intern, and Buck liked her immediately. He told her the full Nancy Story, all three minutes fifty seconds' worth. He also told her about the first traveling team he ever played for. It was called the New York Tigers, though the team was based in Miami and not one person on it had ever actually been to New York. Buck said they traveled through small towns in the South, and people would ask them about jazz and food and Billie Holiday and the poet Langston Hughes and the Harlem Renaissance.

"What did you do?" the reporter asked.

"We'd tell them like we knew," Buck said.

"You mean—"

"We lied."

After that, Buck did another radio interview, this one was based in Atlanta, and then he did another interview for a New York radio station, and then he did another magazine interview, and after that he was no longer sure if the Bronx was up and the Battery down.

"What did you drink?" he was asked by a reporter for the *New York Daily News*.

"Funny you should ask that," Buck said. "I always told people my drink was a gin and tonic. But I'd have the bartender fill it up with water. And then I'd act the fool so people wouldn't know."

"Why didn't you want to drink?"

"I don't know," Buck said. "I always wanted a clear head. I guess I thought that was the best way to survive in this crazy world. I sure could use a drink now, though."

AFTER ALL THE interviews and a day in the city, Buck talked about getting back to his beloved golf course in Kansas City. That was Swope Park. Every clear day, you could find Buck at Swope. He would hit golf balls and chase them around in his golf cart. He was a pretty good player and he was an excellent golf cart driver. But he played golf for another reason. He said:

> *I gotta get back*
> *To Swope Park,*
> *Put my tee in the ground,*
> *Hear Swope say,*

"Where you been, Buck?"
I say, "Been all over."
And Swope will tell me,
"You know better than that.
Come on home."

I had grown used to seeing Buck tired at the ends of days.
Buck was wearier than usual. He had been sucker-punched
by the Star interview and then pounded relentlessly by so
many interviews and requests. His head spun. He was hun-
gry. He was surrounded by a Friday evening in New York—the
construction sounds, the blaring horns, the fast walkers, the
street hustlers, the Broadway lights, the hole in the sky. Buck
loved New York. He was ready to get home.

"I'm going to sleep," he announced when the car pulled up
to the Marriott. As we stepped out of the car, I noticed a woman
standing outside, near a concrete bench. She was wearing a
red dress. It's not quite right to say I noticed her, as if this took
some doing. She was noticeable. Her dress blazed candy-apple
red. You could see it from Brooklyn. The woman who wore it
looked nothing at all like Marilyn Monroe, and yet that was
the name that came to mind. Marilyn. It was that kind of
dress. We walked into the hotel, and I turned back to mention
something to Buck about the woman and her red dress. He
was gone. I looked back to see if he had stayed in the car. But
the car was gone too. I looked down the hall. Empty. Bath-
room? Empty.

Then I looked outside. There was Buck talking to the
woman in the red dress. Buck talked and she laughed, she
talked and he laughed. They hugged. She kissed him. A
young man walked over, and Buck talked to him, they hugged,

they all laughed. The three of them stayed together for a long time, Buck and the woman and the young man. Finally Buck hugged them both and walked in looking fresh as morning. Star was a long way back in his memory. Buck said, "Let's go get something to eat."

As we walked to the restaurant, he asked: "Did you see the woman in the red dress?"

"Yes."

Buck shook his head and looked me in the eyes. And very slowly, with a teacher's edge in his voice, Buck said this: "Son, in this life, you don't ever walk by a red dress."

THESE FOOLISH THINGS
(REMIND ME OF YOU)

In every city in America, at least five times a day, someone asked Buck O'Neil to tell the Nancy Story. Satchel Paige used to call Buck "Nancy." Why? That was the story. Most people who asked for the Nancy Story already knew it. They wanted to hear Buck tell it, because people never tired of it. It was classic. No matter how many times he sang it, Frank Sinatra fans always wanted to hear "Summer Wind" once more.

"Buck, can you tell the Nancy Story for my daughter?"

"Buck, I've heard it before, but can you tell the Nancy Story again?"

"Buck, someone told me to ask: Why did Satchel Paige call you Nancy?"

The atmosphere had to be just right. The Nancy Story had many variations, but in its purest form it was long, a full four minutes when told in loving detail, and Buck had reached the age when he was not often willing to give up that much time for one story. Robert Plant, when asked to sing Led Zeppelin's interminable classic "Stairway to Heaven," would say, "I'm not singing that bloody wedding song." Buck, when asked for Nancy, would sometimes decline by saying, "I'm not sure I'd live all the way to the end."

Every so often, though, the mood struck Buck, listeners were rapt, and the summer wind came blowing in from across the sea. In New York City, across the street from Rockefeller Center, Buck got that look in his eye. He was ready to do the Full Nancy.

"We were playing a baseball game near an Indian reservation in North Dakota," he began.

Buck and his teammates felt different when they arrived in North Dakota. Restaurants served black men in the Dakotas. All of the hotels let them sleep there. Some of the baseball teams in North Dakota had black and white men playing together. "Maybe it was because there were hardly any black people in North Dakota," Buck would say. "Maybe they didn't feel threatened. I don't know." Whatever it was, the Negro Leaguers felt more American in the isolation of Anamoose, Fargo, and Turtle Lake. They were men. Buck often remembered a white boy staring at him across a quiet street in North Dakota. Suddenly, without betraying any emotion, the boy shouted out the word Buck hated more than any other. The boy did not shout it out with malice or anger in his voice, but with a peculiar playfulness, the way children at swimming pools, when prompted by the command "Marco," shouted back "Polo!" The boy looked at Buck with a curious expression. He wanted to know what would happen next.

"Come here, boy," Buck said.

The boy walked over. Buck looked him hard in the eye.

"Why did you say that word?"

"I don't know," the boy said.

"That's a hurtful word."

"I'm sorry."

"Don't be sorry. Don't say it no more."

"Okay."

Then Buck remembered smiling and giving the boy tickets for the baseball game that afternoon. He remembered that the boy's eyes lit up, and he looked up at Buck, and across the years Buck never forgot that gaze of spirit and awe. "He came to the game that night and waved to me," Buck would say. "That boy had probably never seen a black man in all his life. He had just heard someone say that word, and he thought he would say it too. He was a good boy. I'll bet he grew up to be a good man."

That day—or maybe it was another, days do blend together in Buck's stories—a woman came to watch the baseball game in Sioux Falls. In the story, Buck always called her a beautiful Indian maiden. Her name, of course, was Nancy. He never gave details of what Nancy looked like; he expected everybody could see Nancy in their minds.

Satchel Paige was pitching that day, as always. Paige would say he won more than two thousand baseball games in his career, and nobody ever could or would prove different. He won those games with a blazing fastball that catcher Biz Mackey used to say could pound steak into hamburger. He kicked his left leg high when he pitched; Satchel used to claim he kicked so high that his foot blocked the sun. He named his pitches—Bee Ball, Trouble Ball, Midnight Rider, Long Tom, Jump Ball—though they were all variations of the hamburger-pounding fastball theme. Every so often he would stop in the middle of his windup, stand frozen for an instant, and then pitch the ball. This was his hesitation pitch, and it left hitters swinging their bats at shadows. The crowds loved Satchel Paige.

That day Nancy, like everyone else in the stands, watched Satchel Paige. She sat behind the dugout. "Now why did she sit there?" Buck asked. "It was because she knew that Satchel would talk to a dead tree!" Nancy was no dead tree. He asked her to dinner. Nancy said she was babysitting. He told her to bring the kids along.

"We had a good time that night with those kids," Buck said. "We took care of them, me and some of my teammates. I don't think Satchel even knew there were any kids there. He was talking to Nancy."

"You know we're going to Chicago next week," Satchel told Nancy. "You should come along."

"I do have family in Chicago," Nancy said. Satchel gave her the train fare and said he would meet her at the Evans Hotel on the South Side of Chicago. She promised to be there.

Buck said the key to telling the Nancy Story, or any good story, was to tell it slow. Linger. He said people always got in too big a hurry. Don't rush, he said. Savor the details. Follow the turns. Go with the wind. Come to think of it, he said, there is something about life in that wisdom. Buck liked long car rides. He liked plane rides. He always liked getting places.

> *I never minded*
> *Riding the bus*
> *Back in the old days.*
> *Other guys hated those rides.*
> *Complained the whole way.*
> *Said: "We ever going to get there?"*
> *I'd read the paper*

Or talk with somebody
Or just look out the window,
Watch the trees.
We'll get there.
We always get there.

Satchel and Buck sat in the Evans Hotel lobby in Chicago. They sipped tea and looked out the window. It was a bright summer afternoon. They talked about nothing at all. Buck said these are the moments ballplayers miss most. "Isn't that funny?" Buck asked. He said ballplayers, when they think back to their playing days, they don't hold on to the sensation of the big moments, the home runs, the strikeouts, the spraying champagne, the curtain calls, the adoring women, the money. Instead they will remember that a beer never tasted as good as it did in the clubhouse. They will think about those hot afternoons in the bullpen, pitchers spitting sunflower seeds and laughing about some awful little town that smelled like a paper mill. Buck often thought back to that afternoon in Chicago, lemonade, taxicabs, a woman walking by with a pink hat and a big white purse, a man swinging his pocket watch, a warm wind, and that big bay window in the Evans Hotel where Buck and Satch watched the world go by.

A taxi pulled up. Out stepped Nancy. "Pretty as a picture," Buck said. Satch jumped up, rushed outside, and took her arm. He led her inside. "You remember Nancy, don't you, Buck?" Satchel said. The way Buck remembered it, this would be the last time Satchel called him Buck.

"Nice to see you again, Nancy."

Together, Satch and Nancy climbed the stairs. Buck had his tea to finish, and he sat in front of that window for a while.

He watched another taxicab pull up to the Evans Hotel. Out stepped Lahoma. This is the time in the story when Buck pointed out that Lahoma was Satchel Paige's fiancée.

Buck jumped up and rushed outside. He took Lahoma by the arm and said, "Lahoma, it is so great to see you, Satchel has gone off with some reporters, but he will be back presently, why don't you sit here with me, and we will sip some tea until he returns. I will go over and have the bellman take your bags up to the room."

Buck carried Lahoma's bags to the bellman. He whispered, "Hey, man, you better get upstairs and tell Satchel that Lahoma is here. You go ahead and put Nancy and her bags in the empty room next to mine. You got it? And you give me the signal when it's all clear." He stuffed a dollar bill in the bellman's hand—a dollar, naturally, being a lot of money in those days—and Buck returned to Lahoma.

Bellmen in those days, Buck explained, knew what to do.

A few minutes later the bellman walked by and nodded to Buck. That was the signal. After a few more minutes, Satchel walked by the hotel. He saw Lahoma and Buck through the big bay window, and he waved. It was a nice touch. Satchel ran through the front door and gave Lahoma a hug. He said, "Lahoma, it's such a pleasant surprise to see you. I am so happy to have you here in Chicago." Buck wondered how he had gotten outside without being seen. Later, Satchel would explain that he had climbed down the fire escape.

"We had a good time that night. Joe Louis stopped by to say hello. He was great friends with Satchel. Jesse Owens was in town that night, and he came by to say hello too.

"Everything was always happening around Satchel. It was never dull with him. People just wanted to be around him.

He might have been the most famous black man in America then. We loved Joe Louis, of course, and he was more famous among white folks. You had black writers like Langston Hughes, yeah. And you had the musicians, you had Count Basie, yeah, you had Duke Ellington, yeah, Marian Anderson, but I do believe that in that time the biggest hero in the black communities was Satchel Paige."

That night Buck listened through his hotel door at the Evans Hotel. He had not seen Nancy all night, but he knew Satchel would have to come by her room and give her train money to get home. That was ballplayer etiquette. So Buck could not sleep. He was dying to hear what Satchel Paige would say. He listened through the closed door, and at some hour past midnight he heard Satchel's door creak open. *Uh-huh,* Buck thought, *it's going down now.*

Buck heard footsteps tap past his room. He heard Satchel knock softly on the door.

"Nancy," Satch whispered.

No answer. He knocked a little louder.

"Nancy," he said in a plain voice.

No answer still. A louder knock.

"Nancy!" he snapped. And then: *"Nancy!"*

With that last *"Nancy!"* Buck heard Satchel's door creak open again. He knew: That had to be Lahoma. In an instant, Buck found himself racing to his front door. He didn't know what he was going to do, but he knew he would do something. He opened the door.

He said: "Satch, did you want me?"

And Satchel, hitting the beat as always, said: "Yes, Nancy, what time does the game start tomorrow?"

"And," Buck said, "I've been Nancy ever since."

FATHERS AND SONS

Over chicken wings, Buck O'Neil talked about the yellow juice. This was my favorite story of the Negro Leagues. The yellow juice story, of course, involved Satchel Paige. I had asked Buck about it many times, but he had not remembered much. Something about that night— the wings and blue cheese dressing, the aftershock of the Star interview, the buzz of New York City—something triggered his memory.

> *Sometimes I remember*
> *Clear as yesterday*
> *But other times*
> *I have to wait*
> *Until the fog clears.*
> *It's something growing old.*
> *The fog hangs around longer*
> *And you wonder if it will ever blow away.*

Here, in a smoky New York hotel restaurant, the fog lifted and the yellow juice story came back to him like a plane descending from the clouds. It was 1939. Satchel Paige had hurt his arm in Mexico the previous fall. The way Satch told it, the blistering hot food down there angered up his stomach, and the high altitude in Mexico played tricks with his mind. Satch had tried to throw curveballs. Understand, Satchel was a fast-ball pitcher all his life. He had learned how to throw by fling-ing stones at birds. He threw baseballs and stones the same way—hard and straight. They call it "control" in baseball. Satch had control. He threw the ball exactly where he wanted it to go. For show, he would warm up before games by throw-ing baseballs over sticks of chewing gum he had placed near home plate.

The one thing Satchel Paige could not do, though, was make a baseball curve. No matter how he gripped the ball, no matter how much he twisted his wrist, the ball went pretty straight. Hitters mocked him and dared him to throw a curve-ball. Satchel Paige laughed. "Don't need a curveball to get you out," he used to say. And he did not.

Well, way down in Mexico, Satchel Paige decided to show the people he could throw a curveball after all. He contorted his arm, coiled his arm, wrenched his arm, all in an effort to throw curveballs in the light air. The ball did curve, much to Satchel's delight. The crowds cheered. Soon, though, Satchel could not lift his arm over his head. Pain sizzled. One doctor said he was through pitching. The second-opinion doctor agreed. So did the third. A *Kansas City Call* reporter wrote this: "The great one owned a wing that was as dead as a new bride's biscuit."

Kansas City Monarchs owner J. L. Wilkinson signed Paige more or less as a publicity stunt. Wilkie loved publicity stunts. Ten years earlier he had hocked everything he owned and bought a set of portable lights that could be transported on trucks. The Monarchs played night baseball games under those lights all over the country, and this was long before any team in the Major Leagues played night baseball. Those lights made a hideous sound. A million moths fluttered around them. The towers were so short that baseballs routinely flew above them, disappearing into the black sky, and panicked outfielders from opposing teams covered their heads as fly balls crashed down like hail. "We knew the secret," Buck said. "We would follow the ball up into the lights and then run to the spot where the ball was going to land. I remember every time a runner would get a hit and stand on first base, he would ask me: 'How do you guys catch fly balls under these lights?' Of course, I never told them." People came from miles around to see the Monarchs play at night.

Wilkinson told Paige they could be good for each other. He didn't have to sell Paige—nobody else wanted to give him a chance. Paige had infuriated just about everybody else in the Negro Leagues by skipping out on contracts and demanding a huge portion of the gate whenever he pitched. Wilkie was his hope. In the black newspapers, the signing of Paige was hailed as a move that would lead the Monarchs to become the best team in black baseball and perhaps all of baseball. But Wilkinson had no intention of putting Paige on the Monarchs, not with his dead arm. He created a new team instead. It was called the Baby Monarchs at first, though after a while, for publicity purposes, the team became known as the Satchel Paige All-Stars. The All-Stars often outdrew the Monarchs themselves. It was some team. The catcher, Slow Robinson, sang gospels during games. His brother Norman slid with his spikes high. The shortstop, Mex Johnson, quit in the middle of the season to go back home to Texas and teach school. A once-great old ballplayer, George Giles, kept showing up for games, though all he ever did was complain and threaten to quit. Cool Papa Bell played now and again. The team's manager, Newt Joseph, shot rabbits out the window of the bus as it rode along from town to town.

The star attraction, Satchel Paige, threw such soft pitches that for the rest of his life he would talk about one boy in the stands who asked his father: "How'd he ever get anybody out?" Even those soft pitches sparked excruciating pain. Wilkie fixed the games so that most of the opposing hitters swung and missed Satch's slow fastballs. It was a living. Of course, this was Satchel Paige—the man Joe DiMaggio had called the best pitcher he ever faced—and so every once in a while one of those opposing hitters, looking to prove a point, would break the deal and hammer a Satchel Paige pitch to the wall or over it. Buck never knew what made him sadder—the men who hit Satchel Paige's pitches or the men who missed on purpose.

Then, Jewbaby Floyd started to travel with the All-Stars. Nobody knew for sure why they called him Jewbaby. It might have been because he looked old enough to have been one of the Israelites wandering the desert with Moses. Jewbaby was a trainer of sorts. He had been giving massages at the Belleview Motel for years. He also worked with the Monarchs players. He rarely spoke. One day he told Wilkinson, "I can fix Satchel Paige's arm."

There's no telling if Wilkie believed him. But he did send Jewbaby on the road with the team. Sometimes, during games, he became part of the act. Satchel Paige would bend over in apparent pain, and Jewbaby would rush out with a cup of bicarbonate of soda. Satchel would drink it down very slowly while the batter waited. He would then unleash a mighty burp that could be heard all through the stadium. After games, people did not talk about Satchel's disappointing fastball. They talked about his awe-inspiring burp.

Between games, Jewbaby worked over Satch's arm. He poured scalding hot water over Satch's shoulder and then dabbed it with an ice-cold towel. He poured freezing-cold water on the arm and then wrapped it in a boiling-hot bandage. He dug his fingers deep into Satch's shoulder, as if trying to touch the bone, and then he poured more hot water and then more cold. He pulled out his medicine bag, which was filled with bottles and flasks of homemade potions. Some of the potions were dark as oil, some clear as water. Some were thick as tar, others poured easy as scotch. Each day, Jewbaby tried a different potion, and then attacked Satchel Paige's shoulder. More hot water. More cold. On his own, Satchel Paige took to taking showers with water as hot as he could stand.

Nothing worked. Satchel Paige kept throwing pillows. He kept saying the pain was too great for him to go on. He talked about quitting the only thing he knew how to do. That's when Jewbaby Floyd brought out the bright yellow juice. Jewbaby rubbed it into Satchel Paige's shoulder. The juice smelled so ripe, Buck said, that mosquitoes would not get near him.

As soon as Satch started using that yellow juice, he said his arm felt a little stronger. And then a little stronger. And then, finally, he was in Chicago, pitching a game, and he said to his manager, Newt Joseph, "Turn 'em loose tonight." He wanted the other hitters to play it honest.

"You sure?" Joseph asked.

"Yes. Satch is back."

Like that, the fastball was back. Slow Robinson wrote in his book *Catching Dreams* that Satch had a no-hitter going into the eighth inning when, for fun, he lobbed a pitch to his friend Pep Young. "See if you can hit that one, Pep," he shouted with the ball still in the air. Pep hit it off the wall, a triple. He stood at third base and smiled. "I can hit that one, Satch," he said.

"You just sit there," Paige said, " 'cause you ain't going no further."

Satch struck out the next three men. He would pitch for another twenty-five years.

"Of course, I don't think the yellow juice did it," Buck said. "We didn't know anything about arm injuries then. Satch probably just had a dead arm, and he needed rest. But I don't know. People always think there are simple explanations for everything. Sometimes there are miracles. Doctors today want to cut you up. Sometimes you just need a little something that makes you believe."

The waitress brought Buck a salad, and talk shifted to other things, to steroids and politics and women in red dresses. Buck kept waiting for the waitress to come back and fill up his iced tea. Instead a tall man walked up to the table. He looked vaguely familiar. He softly said, "Hi, Buck."

Buck looked up at the man's face, jumped to his feet, and started laughing. He said, "Come here, boy. Come here." They hugged for a long time, each one laughing. Buck stepped back and looked in the man's eyes for a long time and finally said, "My God, you look just like him. You know that, right? You look just like him."

The man nodded. Then Buck turned to us at the table and said, "Do you know who this is? This is Satchel's boy."

ROBERT PAIGE DROVE trucks for Roadway. He told us it was a good life. He had driven more than a million miles without having an accident—that put him in the "Million Mile Club." He was proud of that. He had never had a speeding ticket, which made him very different from his father. Satchel Paige drove fast. Players refused to ride in the car with him. Mex Johnson, the ballplayer and schoolteacher, rode cross-country with Buck one time and was so scared that he finally asked Satch to just drop him off in the middle of the country. "I'll find my own way home," he said. Another time Satch almost missed a Negro Leagues World Series game because he got pulled over for speeding in a small Pennsylvania town. The police officer took him to the judge, who was getting his hair cut. Satchel Paige waited until after the judge had finished. "How do you plead?" the clean-shaved judge asked.

"Guilty, here's my money," Satchel said, and he hopped back into his car and drove twice as fast the rest of the way. He made it to the game by the second inning and pitched the rest of the way.

Robert remembered his father getting pulled over often when he was a kid. One time, Robert was in the car with his mother, father, and sister and they saw the familiar red lights flashing. The officer took Satch's driver's license and told everyone to stay in the car. As he headed back to his police car, Satchel said, "I'll be right back." He stepped out, walked back to the police car, and got in. An hour or more passed. When

Satchel returned, he said, "It's okay, we worked it out," and they drove off, faster than ever. "Give him enough time, he could charm anybody," Robert said.

Robert seemed to enjoy talking about his father here in this hotel bar. He had been called many times by writers to talk about Satchel Paige, and Robert mostly had avoided them. Here, with Buck, it seemed like there was no one he would rather talk about.

"People always get the driving thing wrong," Robert said as he ordered another beer. "My father was not a bad driver. He was a very good driver. He just drove really fast. Whenever my mother would drive, he would say, 'Hold on, let me out, I can make better time walking.' "

WE WERE JUST talking about your dad," Buck said. Robert Paige nodded. He figured people talked about his father all the time. Buck relived the yellow juice story and a couple of others, and Robert said, "Have I ever told you about the keys?"

Buck shook his head.

"I know I have," Robert said, but he went on. He said Satchel was a great collector of junk. He had an astounding assortment of garbage and gadgets scattered around the house— guns that would never fire, weather vanes that had fallen off of rooftops, busted watches, car parts of various shapes and sizes, chains and pulleys from mysterious devices. More than anything, Satchel Paige loved locks and keys. He had piles of locks, and buckets of keys, and he never used any of them. He liked having locks and keys.

One day he handed a lock to Robert and said: "You want to

make a dollar? Go through this bucket and find the key that fits this lock."

"A dollar was a lot of money then," Robert said, because you have to say that, it is required by law, when telling an old story. Robert Paige spent the day digging through the buckets, pulling out keys, and trying to stick them into the lock. For a while he felt certain that none of the keys would fit, that it was all just an intricate gag. His father was always pulling jokes like that. It was one of the many things that made him hard to know. After a couple of hours, though, Robert found the key that opened the lock.

"I found it!" he shouted, and Robert brought the lock and key to his father.

"All right, then," Satchel Paige said, and he looked impressed. He handed his son a sheet of paper. Robert looked at the paper curiously.

"What's this?" Robert asked.

"It's paper," Satchel said. "Go ahead. Make as many dollars as you want."

"That was my father," Robert said. "In time, he gave me the dollar. But not before he scared me into thinking he wouldn't."

ROBERT PAIGE BROUGHT up Dan Bankhead. Baseball fans assume that Satchel Paige was the first black pitcher in the Major Leagues. He was not. That was Dan Bankhead. He was one of five baseball-playing brothers from Empire, Alabama. Branch Rickey, the Brooklyn Dodgers general manager who signed Jackie Robinson, personally scouted Dan Bankhead.

Bankhead threw hard. He threw a great curveball. He had

good control. He was a superb athlete. Rickey called him a sure thing, a certain star, but when Bankhead pitched in his first Major League game against the Pittsburgh Pirates, he gave up eight runs. Rickey promised the doubtful sports reporters that Bankhead would become a great pitcher. He never did. It takes assurance and poise to be a great pitcher. Dan Bankhead never found his balance.

"See, here's what I always heard," Buck said. "Dan was scared to death that he was going to hit a white boy with a pitch. He thought there might be some sort of riot if he did it. Dan was from Alabama just like your father. But Satchel became a man of the world. Dan was always from Alabama, you know what I mean? He heard all those people calling him names, making those threats, and he was scared. He'd seen black men get lynched."

Robert nodded. He had heard the same thing.

"People always hear the story of Jackie Robinson or my father," Robert said, "and it's like they don't realize just how hard it was, how hard it was to be a black man then."

"It was hard for your father too," Buck said. And Buck repeated a story he told often. He said he once went with Satchel Paige to Drum Island near Charleston. They stood on the docks where slaves had been sold after being shipped from Africa. They stood mute for a long time as they looked over the water.

"Nancy, I feel like I've been here before," Satch said.

"Me too, Satch," Buck said. "Me too."

MY FATHER BELIEVED in every superstition you ever heard of," Robert Paige said. He said Satchel believed in lucky clothes, lucky days, and lucky signs. He believed that a man had to be the first to cross through your door on New Year's Day—if a woman entered your house first, it could trigger twelve years of bad luck. Robert could remember a woman standing outside their house on New Year's Day one year, banging on the door, but Satchel Paige would not let her in. Of course, Robert said with a small smile, Satch may have had other reasons than bad luck.

Satch avoided black cats and broken mirrors. He never once in his life walked under a ladder. He believed that if a broom swept over your foot it led to some vague sentence of bad luck, Robert forgot how many years. Satch would never put his hat on his bed, and Robert still shuddered over his father's wrath when he did it by mistake.

More than anything, Robert remembered his father's fear of tornadoes. Nothing scared the man more. This was a geographical problem because Satchel settled in Kansas City, and tornadoes had a long season in Kansas City.

In time, Satchel worked out his fear this way. He got himself a two-sided ax. When the dark clouds came, Satchel sent Robert out into the rain. He ordered Robert to bury that ax in the ground right in the line of where the tornado might head in. It was important to put the ax in precisely the right place. Satchel would lean his head out the open window and direct Robert—a little to the left, no, a little to the right, no, more to the right. And, on Satch's word, Robert would raise the ax high, drive it into the soil, and leave it there until the tornadoes had passed.

"What was that supposed to do?" I asked.

"The ax was supposed to split the tornado in two," he said.

Buck laughed hard. "Split it in two. There was nobody like Satch."

Robert took a long sip of Corona. "No tornado ever got to the house," he said.

BUCK HAD ASKED Robert Paige if he ever thought about becoming a pitcher, and for the first time that night, pain blazed in Robert's eyes. He shook his head, and the topic shifted to other things—to Robert's kids, to the biblical struggles of the Kansas City Royals, to New York traffic, to basketball, to the difficulties of driving a truck with three cars attached.

"I pitched in Little League once," Robert said suddenly. Robert admitted he could throw hard—Robert said "I could throw hard" like it was a curse—but he never wanted to pitch. He never wanted people to point at him on the mound and say, "That's Satchel's boy." He remembered the one day he pitched vividly. It was a hot Kansas City evening. The smell of barbecue lingered in the air. The coach said: "How would you like to pitch today?"

Robert said: "I don't know."

"Let's give it a try."

Robert walked the first batter on four pitches. He walked the second batter too. He walked the third and the fourth and the fifth, and all the while his coach yelled, "It's okay, just relax out there, they're afraid of you, just throw your pitches, they're *afraid* of you." Robert looked over the other team, and he saw that the coach was right, they were afraid of him, but that brought him no comfort. They were afraid he might hurt them. They were afraid because he was Satchel Paige's son. The batters walked reluctantly to the plate and stood with the bat frozen on their shoulders, their eyes wide open, their bodies ready only to spring out of the way if he unleashed a ball at their heads. He walked the sixth batter and the seventh, and he waited for the coach to save him, to get him out of the game. But the coach just watched now with the same horror on his face as the players on the other team. Robert Paige, the son of the greatest control pitcher the game had ever known, could not throw a strike. After he walked the eighth batter, he simply stopped pitching and waited to be taken out. The coach walked out and said "Sorry" as he took the ball. Robert Paige walked back to the dugout with tears in his eyes, and he did not hear a thing.

Robert never pitched again. He played other sports. Robert became a high school basketball star, and then he went into the army, and later he became a truck driver. He did not dislike his life. He said that every once in a while on one of those long drives—not often, but every once in a while—he would wonder what might have happened had he gone back to the mound and tried to pitch again. You think of some crazy things on those long rides.

"When I walked off that mound, Buck, I looked around for him," Robert said. "It's funny. I knew he wasn't at the game. He was never at the games. That never bothered me. He was off pitching somewhere like always. Still I remembered looking for him that day. I looked all over the park. I thought he might be hiding behind a tree or something. That was one of those times I wished he was there. And he wasn't."

Buck nodded. For a minute, Buck looked as if he would say

something. Instead he put his hand on Robert Paige's shoulder and didn't say a word as the waitress finally filled his iced tea.

R O B E R T P A I G E W A S already in the Shea Stadium parking lot in Queens the next morning when the car dropped off Buck O'Neil. Balloons were tied to chairs, and Cracker Jack boxes stacked on picnic tables. Roadway, one of the biggest trucking companies in America, had sponsored a traveling Negro Leagues trailer museum. The Roadway people wanted to do something with baseball—they wanted to show their clients and employees that Roadway shared the best attributes of the national pastime: speed, teamwork, reliability, innovation. The only trouble they found was that baseball in the twenty-first century also seemed to be about high salaries, steroids, competitive imbalance, and labor issues—and those did not quite fit with the Roadway corporate strategy. Roadway wanted to connect to that time when baseball was still played in the daytime, when players played ball for love and all that. So they embraced the Negro Leagues. They had this traveling museum built, called it "Times of Greatness," and they brought it to every Major League city in America. Of course, they brought along Buck O'Neil.

That day in New York, the Negro Leagues museum trailer was open for clients to walk through. There was a batting cage where a machine spit Wiffle balls at Roadway employees. A muffled version of "Bad, Bad Leroy Brown" croaked and wheezed over a busted loudspeaker. It was a party. Reporters waited for Buck.

"Who was the greatest player you ever saw?" the Associated Press reporter asked.

"Well, the greatest Major League player I ever saw was Willie Mays. But you asked about the greatest player I ever saw, and that was Oscar Charleston. He could hit you fifty home runs. Steal you a hundred bases. Old-timers like to say the closest thing to Oscar Charleston we ever saw was Willie Mays."

"What do you think of steroids in baseball?" a television reporter asked.

"As long as people have played baseball, players have tried to get an edge."

"What do you think of this traveling museum?"

"It's outstanding. We're bringing the story to America. That's all I ever wanted."

"Do you wish you could play today with all the money in the game?"

"No, I played at a great time for baseball. Shed no tears for me. I was right on time."

He spoke quickly and softly, without emotion, as if he were reciting the Pledge of Allegiance in a fifth-grade class. He'd done these kinds of interviews so many times. As each familiar question was lobbed at him, Buck offered the same practiced answers. The night before, with Robert Paige, he had been alert, bursting with memories and ideas. Nights like that took their toll. Morning had come. Here, in the parking lot in Queens, Buck was there but not quite, a ghost, an echo. His eyes looked dull and faraway, and he parroted a few old stories, which pleased the people around him no end. But Bob Kendrick looked worried.

"I think we need to get Buck some rest," he said. "He's shutting down."

Bob was right. Buck had stopped talking altogether. He looked over at Bob every so often and nodded, his signal indicating he was ready to go. There were other Negro Leagues players there signing autographs—Jim Robinson, Robert Scott, Lionel Evelyn, Armando Vazquez—and they tried to get Buck talking. Buck was pleasant. But he was not himself.

"Do you have any children?" a man asked.

"No," Buck said. "No, sir."

"That's a shame," the man said.

"Yeah," Buck said.

ANOTHER MAN WALKED over to Buck O'Neil. He carried a huge envelope stuffed with yellowed newspaper clippings. The man said, "Buck, remember me?"

Buck looked up. His eyes showed no flicker of recognition. But he said, "Remember you? Of course, of course. How are ya doing? How's your family?" The men hugged, and the younger man said his family was doing well, he was doing well, very well, he had become a musician, he was doing some things with video and computers. Buck nodded and said, "I'm proud of you, son." By then the line of people wanting autographs had swelled and the younger man walked off.

"He remembered me," the man said. "He's amazing, isn't he?"

"Yes," I said. "I'm sorry, what's your name?"

"I'm Dan Bankhead Jr. My father played for the Brooklyn Dodgers."

Dan Bankhead Jr. I said it was stunning that we had run into him, that just the night before Buck and Robert Paige had been talking at length about his father. "Really?" Dan Jr. asked. He was amazed. He thought nobody ever talked about his father. He wanted to know every detail of the conversation. Then Dan Jr. opened up his clip file, and he slowly flipped through the newspaper clippings, narrating along the way the story of his family. His mother had been a jazz singer; she was beautiful. His father, of course, had been a great pitcher in the Negro Leagues, a hero until he signed with the Brooklyn Dodgers. Then he went to the Minor Leagues, resurfaced again three years later, won nine games, but he was never a hero again. Dan Jr. said he never understood why people did not honor his father more, since he was the first black pitcher in the Major Leagues.

"Nothing against Jackie Robinson, who was a great man, but it was harder to be a black pitcher in those days," Dan said. "People would say, 'Okay, maybe black guys can hit or run, but they can't pitch.' People *still* say that kind of stuff, and it's been, what, fifty years since then?

"I just don't understand why my father has never gotten his due. I want to talk to Buck about that, but I can see he's real busy. I want to see if he can help me tell my father's story. My father was not the best pitcher, obviously. But he was the first. And that story, I think it's an important story. I think it's as important in its own way, you know, as Jackie Robinson. Did you know, my father hit a home run his first time at bat?"

I shook my head. Dan Jr. was talking faster. His words raced his thoughts.

"He was a great athlete. . . . My father wasn't always an easy man to know. He moved down to Houston when I was young, and I didn't see him very much. . . . He was a great pitcher, though, you know, before he was with the Dodgers. Ask Buck about that. Buck will know. My father really blew them away in the Negro Leagues. . . . I can remember one time, he asked me to come down and see him. He told me about how hard it was for him, all the things he had to go through. It was emotional. I remember we both cried. He said, 'They forgot me.' "

Dan stopped. I told him that Buck had said Dan Bankhead would have been a great pitcher in the Major Leagues, but he was scared to pitch fastballs inside against white batters. Dan Jr. nodded. "It's true," he said. Dan Jr. said his father told him that he once had a no-hitter going and then he threw a nice easy pitch down the middle so the guy could get a hit.

"It wasn't time yet for a black man to throw a no-hitter," Dan Bankhead told his son.

Dan said he wanted to do a film about his father. He also wanted to give his newspaper clippings to the Negro Leagues Museum so people could know his father. He wanted people to remember Dan Bankhead. As he spoke, he kept looking back at Buck O'Neil, kept trying to catch Buck's eye. Buck's eyes would not focus.

"He's tired," I explained.

"It's a hot day," Dan said. He stood up and walked over to Buck. "I'm coming to Kansas City so we can talk," Dan said.

"You do that," Buck said. "We need to talk about a lot of things."

"We do," Dan said. "We need to talk about Dad."

Then Dan said he had to go to work. I pointed out Robert Paige, and the men embraced. They spoke for a moment or two about their fathers. Then Dan went one way, Robert went another. A few minutes back, Buck went to the car, and we headed for home.

"Isn't it amazing that we met Dan Bankhead's son one day after you were talking about him?" I asked. Buck looked bewildered.

"Dan's son was there?" he asked. "Why didn't he come up to me?"

I told Buck that he did walk up, and they had hugged, and Buck had said he remembered. Buck nodded. "That was Dan Jr., huh? I didn't know that. I thought he was somebody else." Buck looked out the car window at ugly old Shea Stadium and said, "I sure wish I could remember everything. So much to remember."

SUMMERTIME

Buck kept coming back to the old woman. Every city Buck had visited that summer seemed hotter than the city before, but this was the heat crescendo— Washington roasted. Gnats and flies attacked in the humidity. Buck kept talking about the old woman. She had walked across the street in front of the car, and Buck watched her. She was seventy or so, gray hair, small, she wore a long dress and a silver jacket. She carried two small plastic bags of groceries. She walked slowly, like she was considering whether to turn back.

The woman came upon a puddle in front of the curb, a puddle big enough to have condominiums built around it. Buck said: "Puddles in Washington must be bigger than anywhere else in America. The sewers must be backed up or something." The driver made a crack about sewers and politicians. The woman stood by the puddle a few seconds. She studied it, measured it perhaps. She then seemed to bend her knees and lean forward, as if she intended to jump across. Buck held his breath. Then she shivered, as if a cool breeze had passed through her, and she stood up straight, took one longing glance at the puddle, and made the long trek around.

"Hold on for a second," Buck said to the driver. Buck stepped out of the car and walked up to the woman. He offered to carry her groceries, but she said she could manage fine. He said to her, "I saw you standing there by that puddle."

She smiled. "Yeah?"

"I thought you were going to jump over it for a minute there."

"You did, huh? I thought about it. There was a time, you know."

"I know," Buck said. "There was a time."

"I know exactly how that woman felt," Mamie Johnson said. "We all had our time, didn't we, Buck?" Three Negro League players sat together at a picnic table under a tent that flapped in a hot breeze. This was just outside RFK Stadium, where the Washington Nationals played baseball. While Mamie talked, she signed autographs. She handed out baseball cards of herself. The black-and-white photograph on the cards showed a woman fifty years younger reaching high in the air in a posed effort to catch an invisible baseball. They called her "Peanut" then—Peanut Johnson; she was one of three women who played baseball in the Negro Leagues. That was during the Eisenhower 1950s, and by then most of the best black baseball players played in the Major Leagues. Most of the promising young black players played in the Minor Leagues. The rest of the dreamers played in the Negro Leagues. The players in the Negro Leagues by the mid-1950s were mostly old, flawed, or unlucky—the league was dying. Owners needed stunts and sensations just to draw a few hundred people to the ballpark. Women ballplayers seemed interesting enough. "Mamie could play a little," Buck said.

Across the table, Hubert Simmons tried to sign an autograph.

He had pitched for Baltimore in the Negro Leagues for a short time. His hand shook severely as he tried to finish the final letters on his autograph. He stopped short and apologized. "The shakes go away after a while," he said. The sun scorched the backs of necks. A radio nearby played the song "Kokomo." The disc jockey then said it was ninety-eight degrees. Body temperature. Buck tried to open a bottle of water, but his hands were sweating and he could not twist off the cap.

"Let me help you there, Buck," a fan said, but Buck pulled the bottle away and grunted, "I got it, man." He wrestled with the cap longer. Hubert Simmons gently dabbed his forehead with a handkerchief he'd been carrying in his pocket, and he tried to sign another autograph. His hand shook again. "I'm so sorry," he said softly. Peanut Johnson handed out another baseball card and talked about how the players used to treat her like a younger sister. Buck managed to get the cap off the bottle of water. His face indicated the drink was not worth the effort.

"Were you good?" someone asked Buck.

"I held my own," he said.

"Were you good?" someone asked Buck a moment later.

"I'm getting better every year," he said.

The people kept coming at Buck—some for autographs, some for stories, and some because there was nothing else to do on a hot summer day before a ballgame. Buck's face flushed, and his eyelids drooped, and he said, "Excuse me, I need to do something for a minute." He stood up and walked into the Roadway Negro Leagues traveling museum.

Buck went into the trailer and breathed in the air-conditioning. He looked around at the familiar photographs of Negro Leagues players. He saw a young boy watching a video screen. On the screen, there was black-and-white footage of an old black pitcher named Chet Brewer talking at a dinner. Buck and Brewer had been friends—Brewer died in 1990.

On the video, Brewer told this story:

There was this black youngster who wanted to play for the local white team. He showed up at a game one day, and the manager said, "Get out of here, boy. You know Negroes can't play here." The player showed up the next day, and the manager said the same thing. The next day, the kid showed up again and the manager said, "If you come back tomorrow, I'll call the police."

Well, the player did show up again, and the manager gave in. He gave the kid a uniform, but he told his players, "All right, I know how to get rid of this boy. I'll find just the right situation, send him out there, and embarrass him so much he never comes back." In that game, the bases were loaded, two outs, game on the line, and the manager said, "All right, boy, you get in there and hit."

On the first pitch, the kid hit a long fly ball off the right-field wall. He sprinted around first base. He flew around second. And as he was about to slide into the third base, through the cheering crowd, you could hear that manager scream, "Look at that Cuban run!"

The child in the trailer laughed along with the black and white audience in the video. Buck walked over to him and asked kindly, "Do you know why that's funny, son?" The child looked up, his face slightly red, and he shook his head. Buck said, "I would hope you did not get it, son. See, in those days, in this country, it was better to be Cuban than an American black man. If you were Cuban, you could get served in restaurants.

But if you were black and born right here in the US of A, they wouldn't give you a meal. Isn't that strange?"

The child nodded. Buck said:

Funny,
You look back,
Didn't make no sense.
Racism.
No sense
What people do to each other
'Cause of something dark
In their hearts.

Buck and the child walked around the trailer. Buck pointed to a picture of Oscar Charleston and said, "Do you know who that is, son?" Another head shake. "Oscar Charleston might have been the greatest player this game has ever seen. You know Barry Bonds?" A nod. "Oscar Charleston was like Barry Bonds. He was better than Barry Bonds."

He pointed to a photograph of Cool Papa Bell. A head shake. Buck said: "Cool Papa Bell was so fast that he could steal second and third on the same pitch. That was Cool Papa Bell. Fastest man I ever saw on a baseball field."

He pointed to Josh Gibson. A vague nod. "Best hitter I ever saw," Buck said.

Buck pointed to a stocky pitcher named Hilton Smith. Head shake. "He had the greatest curveball you ever saw. Actually, he threw three or four curveballs. He had a big curveball and a little one. He was something to see."

Buck pointed to Satchel Paige. "I've heard of him," the child said happily.

"He was everything you heard and more, son," Buck said.

So it went. Buck and the child walked around the trailer, they looked at different photographs, and Buck offered a quick commentary on each player. They played a couple of interactive computer games. Then the child pointed to a photograph of a young Buck O'Neil and asked, "Is that you?" A nod. "Were you good?" the child asked.

"It's not right to talk about yourself like that," Buck said. And then he smiled and said that he was a pretty good hitter. "I never did have much power," he said. "I hit those line drives."

He then explained how, over time, he became a much better hitter. When Buck was a young player with the Monarchs, he was often baffled by curveballs and spitballs and emery balls and the various other kinds of junk pitchers in the Negro Leagues would throw. Then he went into the navy during World War II, and he served in Subic Bay, in the Philippines. By day, he loaded and unloaded ships with his all-black battalion. At night, though, he thought about curveballs and how to hit them. He dreamed baseball. He was thirty-five when he got out, but he led the Negro American League in hitting in his first year back, and the next year he fell just one batting-average point short of doing it again.

"You can do anything you set your mind to, son," Buck said.

It was then that I noticed a man watching them both. It was the boy's father. He looked as if he might cry. "Someday," the father said, "he's going to know how much this meant."

But the funny thing was, while the dad was talking, I was not looking at his son. I was looking at Buck. The flushness of his face was gone. His eyes were wide open. He bounced as he walked, and he laughed and talked. When they finished the tour, Buck said, "Well, I've got to go back outside to sign some autographs," and he headed out into the Washington oven and the hungry mosquitoes. He almost ran to the picnic table and announced to all the people wilting in the heat, "You know what? It's a beautiful day. Feels like the sun is on your shoulder." Buck sat down, opened up a bottle of water with one grunt and a twist. He drank half the bottle in one gulp.

GARY, INDIANA

uck O'Neil had to wait forty-five minutes for his ride to show up at Chicago Midway Airport, and it was a good time to take stock. Buck had been home three days all month. He'd spent his time speaking at charity dinners and appearing at schools and signing autographs at ballparks. Now he was going to Gary, Indiana, to appear at something called the Northern League All-Star Game, and even Buck wondered if maybe he was getting too old for all this.

"Why are we here?" Buck asked Bob Kendrick.

"It's a good question, Buck," Bob said.

"It's the only question," Buck said.

After a long while, their driver finally came by. He mumbled some excuses about traffic. Buck noticed that late people always blamed traffic. The drive to Gary took almost two hours. They arrived at the Radisson Star Plaza, where the marquee welcomed the Northern League All-Stars and advertised an upcoming Huey Lewis & the News concert. Palm trees swayed in the air-conditioned breezes of the atrium, and a life-size waterfall spilled blue-tinted water into a swimming pool. Bird sounds chirped. A few people sat outside "The Khaki Club," a bar surrounded by tiki lamps, and admired the waterfall. The woman behind the reservations counter handed Buck a brochure that stated plainly that there was no reason to ever leave the Radisson Star Plaza.

"Mr. O'Neil, you will be staying on Bob Hope Drive," the woman said.

"All right, then," Buck said, and he headed for his room. After getting lost on both Bill Cosby Way and Liza Minnelli Drive, he found his way.

BUCK CHANGED INTO white pinstriped pants, a blue shirt with a tie and blue jacket. His shoes matched his jacket. He then meandered toward the Celebrity Ballroom East, where he would speak at a luncheon for Northern League players and fans. The luncheon would not begin for another hour or more. Buck just had to get out of his hotel room.

The Northern League was what they called an independent league, meaning that, unlike the Minor Leagues, it had no direct connection to Major League baseball. The Northern League featured three types of players. The first group included older players who once showed promise but flamed out in their first shot at the Major Leagues. A handful had reached the Majors for a brief time but for one reason or another fell out. Some of these players felt like they had been cheated out of their destiny; others felt like they cheated themselves by making some critical mistake. They were giving themselves a second shot.

The second group comprised younger players who never got that first shot at the Majors. They were late bloomers or dreamers, men who felt like they had been overlooked or unfairly discounted. They played baseball to be discovered.

Then there was the third group, Buck's favorite group. These were the baseball lifers. They may or may not have had their shot at the big leagues, but that no longer made much difference. The baseball lifers harbored no illusions. They just wanted to play baseball. They needed to play. In the Northern League, they got a couple grand a month, they stayed in cheap hotels, they took interminable bus rides, and they cursed the Fates for giving talent to players who didn't love baseball the way they loved it. But they played because, in the end, they had to play.

Buck loved the baseball lifers. He thought they were the closest thing to the men he had loved and admired in the Negro Leagues.

"This is nice, real nice," Buck said to one of the organizers of the luncheon. The man beamed. He said the chandeliers had been polished. And he said that in addition to salads and baskets of rolls, there was a jar of gummy bears on each table.

"Hey, Buck. Buck O'Neil. We want to sing for you."

A man walked over with a guitar. He looked to be in his fifties, and he was alone. Buck looked at him and then looked around. "Who is we?" he asked.

Three other men materialized.

"What are you going to sing?" Buck asked.

And they began an old gospel, "Joshua Fit the Battle of Jericho." Nobody knows who wrote the words. The arrangement, though, was Elvis Presley's.

Joshua fit the battle of Jericho
Jericho Jericho

Joshua fit the battle of Jericho
And the walls come tumbling down

Buck stepped inside the group, and he started to sing too.

Soon everyone in the room gathered around. The lunch would not begin for another half hour, but people from all over the hotel heard the music and they appeared. They made a circle around the men. Buck danced and sang, he raised up his arms and closed his eyes. All around him, people clapped. Some of them were ballplayers.

"Whoo!" Buck shouted when it ended. His shirt was drenched in sweat. For the first time all day he was smiling. "That sounded like old times."

A MAN WALKED up to Buck in the reception hall and asked, "What Negro Leagues players do you think should be in the Baseball Hall of Fame?" Buck pulled out what looked like an old envelope. It was crumpled, creased, and torn. Names were scribbled on the back, most of them in black ink, the last couple in red. A couple of names on the top were crossed out.

"I made this list a long time ago," Buck said. "I carry it with me wherever I go."

"What is this list?"

"These are the players who I want to get into the Baseball Hall of Fame before I go."

"I see a couple of names are crossed off."

"Yep. We got those in already."

Buck ran his index finger down the list. There were eleven names. He ran them down.

J. L. Wilkinson: "He was the owner of the Monarchs. Great

man. Invented night baseball. Did you know that? We were playing night baseball five years before the Major Leaguers. And Wilkie was one of only two men I knew without prejudice. The other was my father."

Dick Lundy: "Great shortstop. Great, great shortstop."

Double Duty Radcliffe: "He's still alive. He could pitch and catch. Great player."

Biz Mackey: "One of the greatest defensive catchers who ever lived. You heard of Roy Campanella, right? Campy is in the Hall of Fame. Well, Campy said Biz Mackey taught him how to catch."

Newt Allen: "He played second base for the Monarchs for a long time. Had to be twenty years. Great player. I was in awe of him when I came to the Monarchs."

Ted Strong: "Tall, beautiful athlete. He played for the Globetrotters too, back when the Globetrotters were probably the best basketball team in the world. He might have been the greatest athlete in the world at that time."

Bill Wright: "They called him Wild Bill 'cause of the way he would run on the bases. He was wild, man. Great hitter. He moved to Mexico and played ball there because he said they treated him like a man down there."

Mule Suttles: "What a hitter, man. Powerful. He hit the ball a country mile."

John Beckwith: "Never heard of him, have you? He was another great slugger like Mule Suttles. He was mean as a snake too, but that don't mean nothing. Some of the greatest sluggers ever were mean."

Willard Brown: "Could do it all. Hit. Run. Throw. Catch."

Andy Cooper: "My first manager. They say he was the greatest pitcher of his time."

A few months later, five of these men would be voted into the Baseball Hall of Fame. The man asked Buck why his own name wasn't on the list. Buck said: "I wasn't as good a player as any of these men. If I go, it will have to be for something more than my playing. I was a good player, you know, but . . . Anyway it's not my place to say if I belong in the Hall of Fame. What kind of man would I be if I walked around the country with an envelope that had my name on the back?"

WHEN THE MAN walked away, Buck wanted to talk more about John Beckwith.

"He really was mean," Buck said. "They said he used to knock out teammates when they said something to him. Mean as a snake. But you know, that shouldn't have anything to do with the Hall of Fame. I don't like it when people start talking about how a man is off the field.

"I remember when I was on the veterans committee for the Hall of Fame. We were deciding whether Enos Slaughter should go into the Hall of Fame. And I said, 'He should go. Of course.' He was a great player. He hit .300, I don't know how many times. He was a great player.

"And people said to me, 'You can't vote for him, he was a racist.' "

Slaughter was the son of a North Carolina tobacco farmer, and in 1947 he tried to convince his teammates with the St. Louis Cardinals to go on strike when Jackie Robinson joined the Brooklyn Dodgers. This led to a famous statement by National League president Ford Frick. It was, quite possibly, the boldest civil rights statement made in America up to that point.

He directed his statement at the players considering the strike:

"If you [strike] you will be suspended from the league. You will find that the friends you think you have in the press box will not support you, that you will be outcasts. I do not care if half the league strikes. Those who do it will encounter quick retribution. All will be suspended and I don't care if it wrecks the National League for five years. This is the United States of America and one citizen has as much right to play as another. The National League will go down the line with Robinson whatever the consequences. You will find if you go through with your intention that you have been guilty of complete madness."

Later that same year, Enos Slaughter purposely spiked Jackie Robinson.

"In the end, I think Enos probably saw the errors of his ways," Buck said. "I heard that he changed. But that's not the point. The point is, people said to me, 'Naw, naw, you can't vote for Enos Slaughter, he was prejudiced.'

"I said, 'What's that got to do with anything? If we think like that, we won't let anyone in the Hall of Fame. Look around: The Hall of Fame is filled with racists and drunks and all kinds of people. The world is filled with all kinds of people. You can't know what's happening in a man's heart. Could he play or couldn't he play? That's what matters."

IT TURNED OUT that someone at the luncheon had even less interest being in Gary than Buck did. Mike Ditka, the famous football coach, was the main speaker. He sat at the head table and wore what looked like a bowling shirt that had the words "Ditka Classic" on the back. Ditka kept looking at his watch and telling everybody around him that he had a tee time scheduled back in Chicago for later that afternoon. Ditka wanted to make it clear: he was not going to miss his tee time.

Ditka was getting paid quite a bit of money to be there— whispers had the price at $15,000. He was a huge name in the Chicago area. Before Ditka spoke, he handed out trophies to each of the Northern League players. The procession was slower than he liked. Halfway through, Ditka turned to the announcer who was calling the names.

"Hey," Ditka said, "speed it up."

M ike D itka spoke fast. He talked about the cyclist Lance Armstrong. He talked about his own mad passion for the game of football. Of Buck, he said, "He's seen the good, he's seen the bad, and he's seen the good again. This is a great country. But it's not so great that it can't be better."

He then recited a poem of sorts based on a quote from former secretary of agriculture Ezra Taft Benson.

> *Be careful with your thoughts because they turn into words.*
> *Be careful with your words because they turn into actions.*
> *Be careful with your actions because they turn into habits.*
> *Be careful with your habits because they become your character.*
> *Be careful with your character because it defines you.*

Finally he said, "I don't know a lot about a lot. But I know this. You got a dream? Chase it." Then Mike Ditka told everybody about his tee time, and he split.

Buck said, "There's a lot of wisdom in what Mike Ditka said. You just had to listen real quick."

BUCK WAS NOT getting paid; he was there to raise money for the Negro Leagues Museum. Still, he stuck around to do the radio and newspaper interviews. He told the Nancy Story on one radio show in Indianapolis, and he told a young reporter a little story about hitting.

"I really could hit," Buck said. "I didn't hit a lot of home runs—I was a gap hitter. I hit a lot of doubles. And then I became a manager, and I got old, and I couldn't hit anymore.

"But you know what's funny about hitting? You never really stop believing that you can hit. You may not be able to catch up with the fastball, but in your mind you can still hit it just like always. I look out there now sometimes, and I think, *I could hit this guy.* Of course, I couldn't. But that's how you think. I remember, when I was a manager, I couldn't play anymore. But I used to need a pinch hitter. And I would look around the bench at the other players, and I would think, *Hey, I have a better chance of getting a hit than any of these guys.*"

The reporter asked if he ever delivered as a pinch hitter.

"Sometimes I did," Buck said with wonder in his voice. "You know? Sometimes I did."

THEY TOOK BUCK out to the Northern League All-Star Game that evening at U.S. Steel Yard in downtown Gary. It was a pretty little ballpark that overlooked the highway and the Gary Works steel plant. The World Famous Jesse White Tumblers flipped and rolled before the game started. Buck loved anything that had "World Famous" in the name. A local policeman gave a stirring rendition of the national anthem. Fireworks popped and exploded, lighting up the KRAZY KAPLAN'S FIREWORKS sign in the outfield. The players took the field.

"Isn't this wonderful?" Buck asked.

He was a new man. I'd seen him emerge out of funks and exhaustion many times—like Houdini escaping from locked trunks—but he had seemed so run-down earlier in the day, I expected him to stay at the hotel. Now he looked refreshed. He ate hot dogs smeared with mustard, and he hopped from table to table to talk with people. A woman gave him a box of candy, and he hugged her. A man asked him about his playing days, and Buck hugged him. The public-address announcer barked out the license plate for a Buck Celica and asked the man to come to the information desk.

"You have just won a free car wash for winning the 'Dirtiest Car in the Lot' contest," the announcer said, and Buck laughed for a long time.

"I wonder if he will come get his car wash," Buck said. "What a great idea."

Buck had transformed. All that day, he had looked distracted, tired, worn down, and now he watched the baseball game happily, signed autographs. He looked as if he could go all night.

BUCK'S DRIVER WAS a kid named Dwight. He was supposed to take Buck back to the hotel around nine-thirty. But at that time Dwight was nowhere to be found. Numerous people worked their walkie-talkies in an effort to locate Dwight. Finally they reached him, and Dwight raced up the stadium ramps to get to Buck. When he reached Buck, he was breathing hard, and he apologized. He said he had been dealing with mascot issues. Dwight said he had one last thing to do.

He took out his own walkie-talkie and called down to the field. He said: "Hey, guys, I have to take Mr. Buck O'Neil to the hotel. I will not be able to participate in the Great Pizza Race. Over. Is that clear? We need someone else for Pizza Race. Over. We need someone else for the Great Pizza Race. Over?"

Static followed. Dwight shrugged and led the way to the car.

"Who are you in the Great Pizza Race?" Buck asked.

"I'm the pizza man," he said. "It's better than being the pepperoni."

"I can imagine."

"I got promoted this year," Dwight said.

In the car, I asked Buck how he did it, how he recovered from his early depression. He shrugged. "You just do it," he said. He sounded like a magician unwilling to give away his trick.

"Did you get a nap in?" I asked.

"No," he said. "I was going to nap. But I got to reading, and every time I would start to fall asleep, I came across a really interesting part."

We rode through Gary for a while, passing numerous fast-food restaurants, and Buck looked out the window. He had a huge smile on his face. We were back to the beginning. Everywhere we would go, people asked him how he did it, how he overcame bitterness, how he avoided hatred, how he kept going happily along at his age. He had lost his wife, Ora, in

1997. He had lost almost all his friends. He had been to so many funerals.

Buck laughed. He told a story. He and fellow scout Piper Davis were looking for a ballgame in Louisiana. They came upon a baseball field. The lights were on.

"This must be where the game is," Buck said to two men standing by the entrance.

"Oh, yeah," one of the men said. "This is where it is, all right."

They drove in and walked toward the field. Buck never could remember which one of them first realized that something was wrong. At some point, though, they were close enough to the field to notice that every man in the stadium was wearing a white hood. A pig roasted on a barbecue pit. A truck was parked on the pitcher's mound, and a man stood on the flatbed. He was preaching something. He was dressed as the Grand Dragon. "No, Piper, this ain't no ballgame," Buck screamed as quietly as he could, and they ran back, jumped in the car, the tires spun in the parking lot, spitting pebbles, and they were gone. They passed the two men at the entrance, who were on their backs laughing. When they could no longer see the lights of the stadium, Piper and Buck laughed just as hard.

Buck laughed again in the car that night. Whenever he talked about racism, he always talked about it like this, with little stories. He laughed about the night he and Piper Davis almost walked into the middle of a Ku Klux Klan meeting. He smiled thinking about the day a young boy in North Dakota called him the worst name imaginable. And even when a memory of racism cut through him, as in this story he told me years ago, he spoke in verse.

My wife, Ora, wanted a hat.
So she went downtown
To a store called Woolf Brothers
Where black women could shop
But could not eat
At the counter.
Ora was allowed to look at the hats,
Imagine how they might sit
On her head.
But if she touched one,
Touched a hat,
She had to buy it.
A black woman could ruin a hat
By touching it.
So degrading,
So degrading.
That's what racism was.

"Hey, Buck," I said in the car that night.

"What's that?"

"What were you reading that was so interesting? A novel?"

"No, no, nothing like that," Buck said.

"Newspaper?"

"No. See. Well. I was reading the Bible."

"The Bible?" I said it louder than I intended. Buck had talked about running into all those interesting parts that kept him from sleep. I figured by now nothing in the Bible would jolt Buck out of sleep. He shrugged.

"Yeah, I was reading the Bible," he said. "I figured I needed some help."

"What kind of help?"

"Well," he said, "I figured I needed some help after some of the thoughts I was having earlier about being in Gary. I thought I could probably use a little spiritual help."

Buck closed his eyes and smiled happily as the car rumbled back to the hotel.

I'D RATHER HAVE A MEMORY THAN A DREAM

In the last summer of his life, Hilton Smith wrote long letters to the Baseball Hall of Fame in Cooperstown, New York. He wanted in. He was not sure how a man went about getting into the Baseball Hall of Fame. Hilton Smith had never played in the Major Leagues. He had never played in the Minor Leagues. He was by nature a modest man, and it had been more than thirty years since he had talked at length about his days as a baseball player. Some of the men and women at church where Hilton Smith was an elder never knew he played baseball. They never even suspected it.

Hilton sat in the kitchen, swatted gnats at his neck, and wrote letters in the fading evening light of Kansas City summer. He addressed the letters to "To Whom It May Concern," and he began each with the claim that he once threw a curveball that he supposed was the best in the world. Hilton did not like to brag, but they needed to know that he could make the ball curve a dozen different ways—up and down, side to side, whatever he wanted. He had practiced throwing that curveball in his childhood, in a dusty Texas town called Giddings. He threw baseballs against a wooden fence all day, into the night, until he had learned all their secrets. Sometimes he felt like the baseball was his marionette. He could make it come to life.

With that curveball, Hilton almost never lost when he pitched for the Kansas City Monarchs. He included some of his pitching records in the letters—twenty-four wins and one loss one year, twenty-three wins and two losses another—but he knew those records were not written down anywhere, and he was not sure the Hall of Fame people would believe him. So he wrote more about the games and the men with bats he had to make look foolish. Hilton wrote that, had the Fates and timing been kinder, he might have been the first to break the Major League color barrier and not Jackie Robinson. In fact, Hilton wrote, he had been the one to convince Jackie Robinson to play baseball to begin with. He also wrote about the rivalry he had with his teammate Satchel Paige. In the Negro Leagues, they called Hilton Smith "Satchel's Relief." It worked like this: Satchel would start every game. This encouraged people to buy tickets. Then, after Satch pitched three innings, Hilton Smith would come in to pitch the final six. Satchel was the attraction. Hilton was the replacement pitcher. Hilton often pitched better in those final six innings than Satch had pitched in the first three. Few noticed. Hilton Smith hated the name "Satchel's Relief."

Satchel had died in June of the previous year. It was 1983. Satch's death had a strange effect on Hilton. He had not liked Satchel Paige—the two men lived just blocks apart in Kansas City but almost never spoke. Hilton believed he was every bit as good a pitcher, but Satchel had received more credit because he was colorful. Hilton was not colorful. He did not think pitchers should be colorful. No, he did not have much use for Satchel Paige.

Still, when Satchel died, Hilton Smith started writing those letters. He started to talk about being forgotten. He asked his dearest friend, Buck O'Neil, if he would ever get into the Hall of Fame.

"You will get there," Buck said. "You deserve to be there."

"They don't know me," Hilton said.

"We will remind them," Buck said.

At home, Hilton Smith would jump whenever the phone rang. "Who was it?" he would ask his son DeMorris. "Was it the Hall of Fame?"

"No," DeMorris would say. DeMorris had played baseball in the Minor Leagues. He was quiet and proud like his father. The two men had a connection—they understood each other's silences. But DeMorris could not understand his father's sudden obsession with the Hall of Fame. DeMorris would ask: "Why do you care about the Baseball Hall of Fame? What does that mean? You know how great you were. What difference does it make?"

Hilton Smith said softly, "I deserve to be there."

DeMorris understood. His father was dying.

Baseball overflows with myths. Nobody is quite certain how baseball was invented—some say it grew out of the British sport rounders, others say it evolved from cricket, others say it came from ball-and-stick games played in antiquity. People have traced its origins to Europe, Africa, Russia, China, and ancient Egypt. About the only thing we know for certain about baseball's past is that it was not invented by Civil War hero Abner Doubleday in Cooperstown.

That myth, however, proved to be stubborn and powerful. Doubleday remained connected with baseball, though he probably never played the game. The Baseball Hall of Fame was built in Cooperstown, a little town in upstate New York. The redbrick museum is surrounded by a scene right out of 1950s America. Families meander along Main Street. They walk past the post office and Cooperstown General Store, past picket fences and the malt shop. Children lick ice cream cones. Old men sit on park benches and talk about Eisenhower. The supermarket on the other side of town is called "Great American." The only thing that separates the town from being a living Norman Rockwell painting is a little shop on the square called "The Latest Obsession," which sells, well, America's latest obsession. You could imagine the owner one day just got tired of changing the store's stock from disco boots to parachute pants to Cabbage Patch dolls to flannel shirts to *Sex in the City* clothes to ringtones. "Yeah, yeah," she would tell the kids who stocked the shelves every time they tried to explain what obsessions were included in the latest shipment. "Whatever. Just put it on the shelf."

In this Great American setting, the Baseball Hall of Fame feels more like a memorial than a museum. Players inducted into the Hall of Fame often invoked the word "immortality" in their acceptance speeches. And they called the place a "shrine." The Hall of Fame meant even more to those great Negro League players who never got their chance to play in the Major Leagues. Buck said to them the Hall of Fame meant redemption. A Hall of Fame induction was their chance to hear at last after all the years that they were great. They belonged in the gallery with Babe Ruth and Stan Musial, Sandy Koufax and Walter Johnson. To those few Negro Leagues players elected to the Hall of Fame, it was bigger than immortality. It was an apology.

A familiar scene occurred again and again throughout the 1970s and 1980s. Old black men waited by the phone. In Detroit, Turkey Stearnes waited. Buck said he had been nicknamed "Turkey" because of the odd way he ran—his arms flapped, his legs flailed, like a turkey. Funny thing: The way he ran was not Turkey Stearnes's defining oddity. He also talked to his bats. Turkey carried his bats in violin cases, and after games he would sit in his hotel room and tell the bats exactly what he thought of their performances that day. He admonished his smaller bats for being too weak, and he scolded his bigger bats for swinging and missing. A bat that relentlessly swung and missed was sometimes threatened with an ax. Turkey's bats usually hit, though. He played for many years—first for ragtag teams in Detroit and later for championship teams in Kansas City—and the researchers who tally such things have concluded that Turkey hit more home runs than anyone in the Negro Leagues, even Josh Gibson. Turkey Stearnes waited in Detroit for the Hall of Fame.

Willie Wells waited in Texas. They called him "El Diablo" in Mexico—"the Devil"—because he played with white-hot intensity. Nicknames defined the Negro Leagues. Many, like Buck, had more than one nickname. Buck was also called "Cap" and "Foots." That was a sign of respect—getting more than one nickname. But almost everybody had at least one nickname. On a long car ride, Buck and I came up with a children's poem of Negro Leagues nicknames:

Tack Head and Tubby,
Cool Papa and Bubby,
Turkey and Piggy,

Sir Skinny and Ziggy,
Murder Man, Boojum, Two Sides, Double Duty.
Pigmeat, Hoghead, Police Car Lopez, and Hootie.
Willie Wells was the Devil. Kissing Bug Rose.
One named his roommate Popsicle Toes.
Cannonball and Schoolboy, Toothpick and Duke,
House Lady, Shack Pappy, Pastor and Fluke.
Cream and Sugar played together,
Sam Jethroe was the Jet.
Everyone called George Sweatt "Never"
'Cause he'd never sweat.
Satan Taylor's mother believed in God,
That's what the papers said,
So they stopped calling him Satan,
Called him Jelly Taylor instead.

The Devil, Willie Wells, slid with his spikes high, and he crashed into pitchers covering first base. So many people threw baseballs at his head that he wore a construction helmet when he went to hit. He was probably the first professional baseball player to wear a batting helmet. As a shortstop, the Devil played like a dream. He chased down so many ground balls that his Negro Leagues teammate Judy Johnson swore he wore roller skates. They also called him "the Shakespeare of Shortstops," and in his later years he wondered when the Hall of Fame would finally call his name.

Leon Day waited in Baltimore for the Hall of Fame call. He had been a great pitcher—he won every game he pitched in 1937. He threw a no-hitter his first game after returning from World War II. In Puerto Rico, according to legend, he struck out nineteen men in a game. Between starts, Day played every position but catcher, and he once ran the 100-yard dash in ten seconds flat even while wearing a complete baseball uniform and spikes. After he finished playing baseball, he became a bartender. People asked him to talk about his playing days, but he almost never did. Leon Day too waited for the Hall of Fame to call him.

All three men would eventually get into the Hall of Fame, in large part because of the unyielding effort of Buck O'Neil. He talked them up, promoted them, politicked for them. He got them all in. None of them, however, lived to see Induction Day. Turkey and the Devil died many years before the Hall of Fame called. Leon Day was alive when Buck O'Neil called to tell him he was going into the Hall of Fame. "He was so happy," Buck said. Leon Day died six days later.

"I never heard that business about Hilton Smith writing letters to the Hall of Fame," Buck O'Neil said as we walked through the Atlanta airport. "Who told you that?"

"His son."

"Well, I was with the Hall of Fame veterans committee at the time. So I would know if that was true. I don't believe that is true. We talked about the Hall of Fame a few times before he died. I do believe he wanted to go into the Baseball Hall of Fame. Anyone who played the game wants that. But I don't believe he wrote any letters."

Buck shook his head as we got out of the airport train and headed to the escalator. Buck and Hilton were best friends most of their lives. They had played together, they had scouted together, and at the end of Hilton Smith's life they talked a lot about what had been. Hilton had grown bitter. Buck hated to see that happen to his friend. He never liked talking about it.

The escalator was not working. Bob said that we should just walk around back; there was an elevator on the other side. "No," Buck said, "we can take the stairs."

The escalator turned out to be much longer than Buck had anticipated. There were fifty or sixty steps. About ten steps in, Buck realized he had made a big mistake. He looked up, counted the remaining steps and realized that there was no chance he could make it to the top. He then looked back down, and saw other people stepping on the escalator.

"Don't look down," he whispered, and he gritted his teeth. He climbed. He started to breathe heavy after a few more steps. "We halfway yet?" he asked in a voice that was meant to be funny but came across as a bit desperate. We were not even close to halfway.

"We're almost halfway, Buck," Bob panted.

Buck climbed. And he climbed. After a few more steps he was panting, his mouth was open, and he sucked air and clutched the railing. Still he shook off the helping hand from behind him. "We're almost there," he said. There were still twenty steps to go. He muttered, "We're gaining on it. We're definitely gaining on it." There were fifteen steps to go. A man walking up behind him said, "Just a few more steps. Just a few more, sir."

"A few more steps," Buck said. There were ten left. Sweat beaded on his forehead. His grip tightened on the railing. You could see the blood rushing to his fingertips. There were eight steps to go. There were seven. There were six. His pace quickened. A smile formed on his lips. "Send a rope for me," he told people standing on top. But he was taking the steps more quickly. He knew he could make it now. Three steps to go. Two. One.

"I'm an old fool," he gasped as he stepped on the top. He walked over to a chair and fell in it. He asked if we could see his heart beating through his chest.

"I've made a lot of big mistakes in my life," he said. "And that right there was one of them."

While he sat, a man walked up to him for an autograph. Buck looked up at him with his you-have-got-to-be-kidding-me expression. But he signed the autograph. The man said, "Buck, are you going into the Hall of Fame this year?"

"I don't know," Buck said. "That's what they're saying."

"I'm pulling for you," the man said. Buck nodded and concentrated again on his breathing. The Baseball Hall of Fame had just announced that a special committee was put together to elect some Negro Leagues players and executives into the Hall of Fame. Buck was not on the committee, which seemed a pretty good hint that they wanted to honor him.

"I'll tell you about Hilton Smith," Buck said between gasps. "Before he died, I promised him that I would do everything I could to get him into the Hall of Fame someday. I thought it was my duty, you know. It wasn't because he was my friend. I was the one who saw Hilton pitch. I know how good he was. He was one of the greatest pitchers who ever lived.

"Well, of course, you know he died before I could help get him into the Hall of Fame. And I remember, at the funeral, I told Hilton, I said, 'I'm still going to get you into the Hall of Fame someday.' It was almost twenty years later, and it was the last year of the old veterans committee, and one of the other guys said, 'Buck, you're always talking about Hilton Smith. How good was he?' And I told them he was like a Bobby Feller. He was like a Greg Maddux. Yeah. And they voted him into the Hall of Fame."

Buck stood up. He had caught his breath. He started to walk again.

"I came back to Kansas City after that, and I went out to see Hilton Smith's gravestone. I said, 'Hilton, you're a Hall of Famer.'"

> *I wish he had lived*
> *To see that day.*
> *He was a little sad*
> *At the end of his life,*
> *Just a little.*
> *Thought he had been forgotten.*
> *I wish he had lived*
> *To hear them say,*
> *"You were great.*
> *You were one of the greatest ever."*
> *But I'm sure he heard it.*

"I really wish he had lived," Buck said. "I miss him. He was my friend."

"You know, his son says he never got back any of those letters," I said to Buck.

"Maybe he wrote the letters but never mailed them," Buck said. He seemed to consider that for a moment. Then he picked up his pace a little, the gate was in sight, and I asked what Hilton Smith would have said after watching Buck struggle to climb the stairs.

"He would have said 'Are you crazy? You're an old man,'" Buck said. "But when I reached the top, he would have said 'Well, I see you made it again.'"

CLASSROOMS IN ATLANTA

There are few absolute truths in the world, but this is one: Pilots are at their most chipper on early-morning flights. This morning, the Southwest pilot decided to play his harmonica over the plane's speakers before takeoff. This has the odd effect of annoying half the people on the plane ("Fly the plane, Bob Dylan," a guy in a window seat mumbled) while enchanting the other half. Buck O'Neil was not in either camp. He was in deep sleep.

He was asleep before most people got on the plane. Buck was a good plane sleeper. For one thing, he was always the first person on the plane. Most of the time, he was bullheaded about his age—climbing those escalator stairs was a typical Buck decision. A woman in the Houston airport once offered him a wheelchair, and she tasted Buck O'Neil's wrath. "I don't need no wheelchair!" he shouted. When getting on planes, though, Buck flaunted his age like a teenager with a new driver's license. "I'm almost ninety-four years old," he announced, triumph singing in his voice. "I deserve to get on the plane first."

Once on the plane, Buck would take out the in-flight magazine and place it on his lap as if he intended to read it. Then he leaned his head forward. That was it. He was out. People would walk by and say, "Hi, Buck!" or "That's Buck O'Neil," but his head never bobbed. Buck often talked about sleeping fitfully in beds. On planes, though, he could sleep through the sound of a hundred harmonicas.

IN ATLANTA, EVERY other street is named Peachtree. Directions sounded like an Abbott and Costello skit. Follow Peachtree Drive until you get to Peachtree Lane, go south on Peachtree Court and you will run straight into the Avenue of Peachtree. From there, you'll want to get on Peachtree . . .

Our driver was lost. Buck was supposed to speak at Daniel H. Stanton Elementary School, and we had traversed enough Peachtrees to reach Martin Luther King Boulevard. "What was the name of this street before Martin was killed?" Buck asked. The driver did not know. It was becoming clear based on the growing number of U-turns that the driver did not know much at all about Atlanta geography. It turned out he did not want Martin Luther King Boulevard, he wanted Martin Street, and then it turned out he went to the wrong Martin Street, and all the while Buck remembered playing baseball in Atlanta. "The white team was called the Atlanta Crackers," he said. "And so the Negro Leagues team was called the Atlanta Black Crackers. You ever heard of a black cracker?"

He remembered playing ball in Ponce de Leon Park, where the Black Crackers played. It was like any other baseball stadium except for one thing: A large magnolia tree towered in center field. The tree stood 460 feet from home plate. The Atlanta legend is that only Babe Ruth and Eddie Mathews ever hit baseballs over that tree at Ponce de Leon Park.

"You know, if you were an outfielder, that tree was your friend," Buck said. "You would play in front of the tree, and when a ball went over your head you hoped it hit the tree. If it hit the tree, you might hold the hitter to a triple. But if it went by the tree, it was a home run all the way."

I said: "Don't you think it's weird that there was a tree in the outfield?"

"Now that you mention it," Buck said, "I guess it was weird. I never thought about it."

They tore down the stadium at Ponce de Leon Park. The magnolia tree still stood in Atlanta.

B UCK WAS LATE to the school, of course, and the program had already begun by the time the car pulled up. Buck walked into the library, and there were a couple of Negro Leagues players sitting on metal folding chairs. Teachers stood with their backs to the walls. And the children were on the floor. The principal reminded them over and over that they were supposed to be sitting crisscross-applesauce. Some were. Others stretched out. Some faced the wrong way and some whispered to each other. Some yawned as Buck O'Neil walked in. How many libraries just like this one had Buck walked into through the years? How many children had he seen? He acted like it was the first time. He sat down in a metal chair and waited for his introduction.

"We are very lucky today," the principal said. "This is Mr. Buck O'Neil."

"Hi, Buck!" the children said.

"Do you kids know how to sing?" Buck asked as he stood up. A few muttered "Yes."

"I don't think you heard me," Buck said. "Do you kids know how to sing?"

The "Yes" was louder now.

"Do you kids really know how to sing?"

The "Yes" was shouted now.

"All right, then. Let's see if you can really sing. I need your help. Everybody hold hands." There was giggling, pushing, whining, but Buck sorted through all of that. After a few moments all the kids and teachers held hands. And he began singing.

> *The greatest thing*
> *In all my life*
> *Is loving you.*

The kids sang softly at first, following Buck on each phrase. This was part of a gospel song called "The Greatest Thing," written by Mark Pendergrass in 1977. The song has other verses, but Buck had been using just this one for so long, it had become a part of him. It was his song. He has sung it at dinners and club meetings and charity functions all over. He always had people join hands.

> *The greatest thing*
> *In all my life*
> *Is loving you.*

Buck held his hand out to one of the teachers as he sang "loving you," and she blushed, sparking giggles among the kids. The children were singing a little louder now, with a little more joy in their voices.

> *The greatest thing*
> *In all my life*
> *Is loving you.*

The kids sang loud now, realizing that this song wasn't going anywhere else. These were the only ten words, and all ten could be understood by three-year-olds. There was something hypnotic about Buck, the way he waved his arms like a conductor, the way he picked up eye contact, the way he pointed at the children not singing with enough feeling. The simple words took on more meaning.

> *The greatest thing*
> *In all my life*
> *Is lov-ing you.*

They sang with him until the end. The children were wide awake. They cheered Buck and themselves, and Buck shouted out, "You can sing! What do you know? I've never heard such singing in all my life." The kids cheered themselves again. Buck smiled at Red Moore and James Lee, the two Negro Leagues players who were sitting behind him, and he said, "A little music wakes up the soul, doesn't it?"

BUCK BEGAN HIS talk. "I'm from the South, children," he said. "I was born so far south that if I had taken one step backward, I would have been a foreigner."

The children giggled again, and he told them about that hot day he worked in the celery fields in Florida. He told them about the day when he discovered black children in Sarasota could not attend high school. Some of the kids listened. And some drifted. The simple reality: No one can keep the attention of elementary-school kids forever, especially when talking about a past they cannot imagine and injustices they have trouble believing.

"Listen, children," he said. "I remember they built Sarasota High School, and I went around singing 'I'm going to Sarasota High School! I'm going to Sarasota High School!' And one day my grandmother says to me, 'Son, you're not going to Sarasota High School.'

"And I said, 'Why not, Grandma?'

"She said: 'Sarasota High School is for white children.' I started to cry, and she said, 'Don't cry, John. I won't live to see it, but someday black children and white children will all go to Sarasota High School.' And wouldn't you know it? A few years ago, they called me back to Sarasota High School to give me my diploma."

The teachers smiled and cheered. The children, many of them, stretched and yawned again.

BUCK TOLD A few more stories and ended by telling the kids: "Don't let anyone tell you that you can't do something. You can do anything and you can be anything you want in this world, children. Remember that."

Applause. The principal asked if anyone had questions. The kids waved their hands wildly. Some of the wildest arm-wavers could not think of a question when called upon. One asked

Red Moore, "What was the hardest part of playing in the Ne-
gro Leagues?" Red Moore was a first baseman in the late 1930s.
He played defense with style. He would catch throws behind
his back during infield practice. "You listen to him, children,
because this is Red Moore and he could really play," Buck said.
"He could really pick it. That's what we say in baseball. He
could pick it." Buck looked over at Red Moore, who nodded
and smiled. Red was almost eighty-nine years old.

Moore said that he remembered restaurants would not let
him eat. He said that was the hardest part. They were hungry.
James Lee nodded. "We ate luncheon meat on the buses,"
Lee told the kids. "We ate crackers on the buses. Sometimes
we didn't eat at all. It was hard. We were treated like we
were . . . less than men."

He said the last words softly, as if he regretted saying them
even while speaking them. Buck looked out at the kids, many
of whom were still waving their arms, hoping to be called
upon. He stood up. And he said, "I can see the way you look.
Oh yeah. You listened to that and you think that the Negro
Leagues was inferior. The Negro Leagues was not inferior."

He stomped his foot. The kids stopped waving their arms.
Buck's eyes looked bloodshot.

"Don't feel sorry for us," he said, now so softly that only the
children in the front row could hear. "We had a great time."

THE PRINCIPAL SAID there was time for one more
question and a little boy in the front row wearing an army
fatigue shirt asked Buck why it was that blacks and whites
hated each other. Buck looked deep into the boy's eyes. Buck
called him "son."

Son, there never was a time
Everybody hated everybody.
Never so.
Always good white folk.
Always good black folk.
Remember, son.
Don't let hate fill your heart.
Always more good people
Than bad
In this world.

THE CAR CIRCLED back and crisscrossed Atlanta Peachtrees, until we somehow reached Turner Field, the ballpark where the Atlanta Braves played. In front were giant numbers with baseballs on them. These were the uniform numbers the Atlanta Braves had retired. Buck did love uniform numbers. He tried to guess the player behind each number.

21: "Well, let's see here, that's gotta be Warren Spahn."

Right.

"Great pitcher. Great man. Warren Spahn. Threw the screwball. Pitchers don't throw screwballs anymore—it's too hard on the arm. Everybody's worried about getting hurt. That's the money. Players are better now than they were in my day. They're bigger and stronger and faster. But we would do anything to get you out. There was no money. We played to win."

41: "I don't know, who is that?"

Eddie Mathews.

"Oh, sure, great hitter. Lots of power. He hit a ball over the tree at Ponce de Leon Park."

44: "Oh, well, that's Hank Aaron, of course. I remember the first time I saw Henry Aaron. He was playing shortstop for the Indianapolis Clowns. He looked like a little kid. He couldn't have weighed a hundred and fifty pounds. I asked their manager, Piggy Barnes, 'Who's the kid?' He just smiled and said, 'You'll see.'

"So this kid comes up, and I tell our pitcher to throw his best fastball. The kid bangs a ball off the right-field wall. Kid comes up again, and I tell our next pitcher to throw *his* best fastball, and the kid hits it off the center-field wall. I tell our last pitcher to throw him a curve, and he hits it off the left-field wall. All the time, Piggy Barnes is in the dugout just smiling.

"So after the game, I go see Piggy, and he's telling me about this kid Aaron, how good he's going to be, and this time I started smiling. He said, 'What are *you* smiling about?' I said, 'Piggy, when you guys come to Kansas City later this year, Henry Aaron won't be on your team.' Sure enough, he was signed by the Milwaukee Braves days later."

35: "I don't know that number."

"It's Phil Niekro."

"Oh, of course, of course. Threw the knuckleball. Outstanding pitcher."

3: "That must be Dale Murphy. He's an outstanding man."

42: "And, of course, that's Jackie Robinson. I was so proud when baseball retired his number for every team. That's the way it should be. Jackie Robinson belongs to the world."

BUCK WAS TAKEN to a conference room where he and a half dozen Negro Leagues players sat until the game began.

The Atlanta Braves were having a special Negro Leagues celebration. More Major League teams were dedicating days to honor those who played from the Negro Leagues, and they all invited Buck to participate. He loved being around his old friends.

There was a certain rhythm to conversations when Negro Leagues players got together. They hugged and almost immediately asked each other who died since the last time they had seen each other. The numbers of Negro Leagues players dwindled—there were, at that time, fewer than fifty players left who played in the Negro Leagues before Jackie Robinson crossed the color line. The number seemed to shrink almost daily.

"You look good," Buck said to an old catcher named Otha Bailey.

"You look good," Bailey said.

"What do you know?"

"Nothing. How's Double Duty?"

"I don't think he has much more time. What about Patterson?"

"Which Patterson?"

"Pat. Did he die?"

"Yes," Bailey says.

"Did he die at home?"

"Yes, at home."

"Good," Buck said. "Well, that's good."

OTHA BAILEY WAS a small man, no taller than five foot six, but even at seventy-five years old you could see the rough and intense catcher he had been in the Negro Leagues many years before. He once missed a chance to try out for the Los Angeles Dodgers. He never explained exactly why he missed it, but when asked if he felt badly about it, Bailey said he did not need to prove himself by playing in the Major Leagues.

Buck heard a story about Otha Bailey once. Buck said there was a big and nasty pitcher named Bill Powell on the Birmingham Black Barons. One game, Powell could not or would not throw strikes—you never could tell with Powell. This wildness went on for an inning or two, and then Otha Bailey, the catcher that day, walked out to the mound. He took off his mask, looked up into Powell's face—Bailey had to look way up—and he snarled: "If you don't start throwing the ball over the plate, I'm going to kick you from one side of this park to the other."

"What happened then?" James Lee asked.

"Way I heard it," Buck said, "Bill Powell started throwing strikes. How about it, Otha? Is that true?"

Otha Bailey looked a bit sheepish. But he nodded.

"Baseball turned me into a madman," he said, "but I wasn't scared of nobody."

JOHN SCHUERHOLZ, ONE of the most admired men working in baseball, walked in. He wore his trademark suspenders. He hugged Buck. They had known each other for almost forty years. One thing that became clear the longer I traveled with Buck: Everybody knew him. Buck and Schuerholz had first met in Kansas City, when Schuerholz worked for the Royals. In time, Schuerholz would become general manager of the Royals, and the team would win the World

Series. Now Schuerholz was the general manager of the Atlanta Braves, and his team had won thirteen consecutive division championships, a record. His Braves, against predictions, were winning the division again.

"I see you're doing it again, John," Buck said.

"It's not me, you know that, Buck."

"You're a smart man, John. A smart man."

"You taught me well."

"Yes," Buck said. "Yes I did."

In time, baseball people always start talking about golf. Schuerholz asked Buck about shooting his age—that's a big thing in golf, shooting a score equal to or lower than your age. Buck first shot his age when he was seventy-five years old, and he has shot his age almost every year since.

"You still shoot your age?" Schuerholz asked.

"Yeah," Buck said. "But that's not a good score anymore."

AFTER THOSE NEGRO Leagues players talked about health and family and the friends who were gone, there was silence. Buck never liked silence. He pounded his fist on the table and asked the men, "Okay, where was the last game that you played?"

That sparked the conversation. There were two distinct eras of players. There were the men who played after Jackie Robinson, and they remembered hard times. The pay was not steady. The owners were not reliable. The buses broke down. Shattered glass scattered in the infields. Some of these players celebrated their pain. They argued about which infield had the most glass and which owner was the stingiest. They debated which town was most racist. They tried to top each other when talking about the worst thing a fan ever yelled at them. They laughed at the pain the way veteran soldiers might joke about basic training.

Buck and Red Moore did not speak about injustice. They alone at the table had played in a different Negro Leagues, the one before Jackie Robinson crossed. That was a league fighting for survival. Hopelessness hovered over their games. The players knew, no matter what happened, no matter how well they played, no matter how many home runs they hit, and no matter how many batters they struck out, they were fated to play in the shadows. That hopelessness was too much for some. In the early parts of the twentieth century, a gifted dark-skinned Cuban player named Luis Bustamante committed suicide. In his farewell note he wrote the five haunting words that summed up crushed dreams and suppressed rage.

He wrote: "They won't let us prove."

But, Buck said, there was something else in the hopelessness. There was freedom. These Negro Leaguers were not playing the game for anyone else. They played only for themselves, the pay, though unsteady, the cheers of their neighbors, and the bliss of playing the game just right. Baseball in the Negro Leagues was a little bit rougher, a little bit sweeter, a little bit faster, a little bit cooler, and a little bit more fun than anything Buck ever saw in Major League baseball. No, Buck and Red Moore did not tell stories about injustice like the others. They told a funny story about Cocaína García, who pitched in Cuba and had a fastball that would numb you like cocaine. They talked about Goose Tatum, the former Globetrotters player, who used to leave the crowd laughing with his baseball clowning but was so mean he once pulled out a screwdriver and stabbed Hilton Smith during a game. Buck told a tale about Frank Duncan. He was a good player and manager, and his wife, Julia Lee, was a renowned jazz singer, a favorite of President Harry Truman. "Baseball and jazz, the two greatest things in the world," Buck said.

In Kansas City, there was a small hole in the back of the dugout. And baseball players were baseball players. They would look through that hole during games so they could see up women's dresses. Sometimes you would see them pushing each other out of the way so they could get their turn to look. Nobody took longer turns than Frank Duncan.

One day Frank Duncan was looking through the hole and laughing. Then, all of a sudden, he stopped laughing. You could hear his shout echoing throughout the stadium. *"Julia!"* he bellowed. *"Pull down your damn dress!"*

> *Funny.*
> *That's what I remember most,*
> *Stories.*
> *Don't remember the games much.*
> *Don't remember names much.*
> *Don't remember the bad times.*
> *I forget who won and lost most of the time.*
> *Stories.*
> *Silly stories.*
> *I remember those.*

RED MOORE'S WIFE sat next to him and buttoned his jersey. Buck watched them for a long time.

"I see you're still taking care of him," Buck said softly when she had finished. He blinked and looked away.

BATTING PRACTICE WAS about to begin, and a woman who worked for the Braves came into the conference room and told the men she was ready to take them down to the field. "First," she said, "if anybody needs to go to the bathroom, then you better go now." It sounded strange, this young woman telling men who had lived hundreds of years to go to the bathroom. Half of them went.

When they got to the field, Major League players rushed over to hug Buck. One of the huggers was Julio Franco, the oldest player in the Major Leagues. Franco had just turned forty-seven, making him the oldest everyday player in the big leagues since World War II had ended. Buck smiled and said, "I could have sworn I played with you in the Negro Leagues, Julio."

Franco could not stop laughing.

AFTER THE CEREMONY, the other Negro Leagues players scattered, and Buck went to a suite and sat alone so he could watch the game quietly. Every so often, someone wandered up to his table and asked him for an autograph or a story. He obliged. Buck looked at peace. He said: "This was a good day."

He looked around the room and said, "Is there anything sweet in this suite?"

The young woman who worked for the Braves walked over, gave her most alluring smile, and said, "What did you have in mind, Buck?"

He smiled and said, "Oh, Lord, I want something that won't kill me. How about a cookie?"

Buck got four or five cookies. He ate one and wrapped the others. He headed for the exit. As he passed the young woman who worked for the Braves, Buck said, "I'm sure I will see you again."

"I love the way you say that," she said. "You are so filled with hope."

And Buck asked: "What else is there in the world?"

ISN'T THIS A LOVELY DAY?

Buck O'Neil had never been to a horse race before, and he looked thrilled as the car glided slowly past the San Diego palm trees and the beautiful joggers. "In my neighborhood in Kansas City, you need a fat belly to go jogging," Buck said. "Not here. Why are these skinny people running?" Autumn was closing in, though you would never know it in San Diego. Seasons blended together here. We were going to Del Mar. Our host was Bob Turnauckas, the president of a company called "Beyond Meetings & Incentives," and he could not stop talking about how much Buck would love Del Mar. He said that it was the track of the stars, it was where Pat O'Brien bet, Sinatra was there. Bing Crosby sang Del Mar's song: "Where the Turf Meets the Surf."

"A lot of that is gone," Bob said with sadness in his voice. "But there's still some star power. You'll see. There's still some glamour at Del Mar."

"All right," Buck said. "I love glamour."

"Look in your envelope, Buck. I included a thirty-dollar ticket in there for you to use."

"All right, then. I'm going to bet on some horses."

Sometimes on our trip—not often, but sometimes—Buck

seemed to shed a few years. It was quite an astonishing thing.
Wrinkles in his face would straighten. His hair would darken.
His voice—so much of Buck's energy was in that voice—would
get louder. In that car, he looked like a kid about to do some-
thing naughty. Buck liked to say that when he played for the
Kansas City Monarchs and especially when he managed the
Monarchs, the team had class. They had suits fitted at Mat-
law's. Bartenders throughout the Midwest were warned to cut
off Kansas City Monarchs players before they drank too much.
The Monarchs, in Buck's memory, stayed sober, played bridge,
and dutifully wrote letters home to their wives every night.
And, of course, they never gambled. Kansas City was a gam-
bling haven in the 1930s. Buck insisted the Monarchs rose
above temptation and craps tables.

"No sir, it meant something to be a Monarch," Buck said.
"The Memphis Red Sox, that team, well, they shot dice."

He laughed. You could tell he was proud to remember the
Monarchs being all class. And you could tell that, in some small
way, he suspected that the Memphis Red Sox might have had
more fun. These were complicated memories. In any case, he
looked forward to betting on horses.

"Nothing but winners today, Buck," Bob said as the car
pulled into a parking garage with flashy cars.

"Nothing but winners," Buck said.

Before the first race, talk swirled around baseball. Double
Duty Radcliffe was on his mind. Duty had played with Buck
in the Negro Leagues. He died the day before. He was 103.
"You're not allowed to feel sad for a man who lived to be one
hundred and three," Buck said. He remembered a Double
Duty story. It seems that Duty—like Buck—had not been to a

racetrack until late in his life. Once he got there, though, he made up for lost time. He bet liberally. In one race, he got a tip and wagered way too much on a horse Buck thought was called "Wild Willow." Duty's face was flushed from the start of the race, but his friends insisted that the tip looked good. Wild Willow took the lead in the backstretch and pulled away. Wild Willow headed into the stretch ahead by three or four lengths and was looking good.

Then Wild Willow started to fade. Duty did not understand the inner workings of horses, and he did not know what furlongs meant. He did not appreciate any of the complexities of the sport of kings. But he instinctively knew a dying horse. He jumped up and down as if he were riding himself. The other horses closed fast. Duty kept shouting and jumping, he prayed and swore and made promises he could not have kept. And finally, when he could not take it anymore, when Wild Willow and another horse raced stride for stride to the finish, Duty ran down to the rail and yelled with everything he had: *"Slide, dammit, slide!"*

"He lost, of course," Buck said. The thought of losing inspired Buck to talk about his favorite team, the Kansas City Royals, who were in the middle of a long losing streak. The streak would end up being nineteen games, the longest losing streak in almost twenty years. But it was not the losing that consumed Buck. The Royals lost stupid. In one game, the manager—a lifelong baseball man named Buddy Bell—had sent out the wrong lineup card. In another game, the Royals led by five runs going into the ninth inning and lost the game after their right fielder drop-kicked a ball and their left fielder dropped a pop-up. One first baseman had hurt

himself swinging a bat during a rain delay. Another first base-
man had been consigned to the Minor Leagues after com-
mitting numerous humorous baseball crimes, like the time
he got hit in the back by a throw from his own outfielder.

The most spectacular play of the Royals season came later,
during another loss, when two Royals outfielders settled un-
der a fly ball. They looked at each other, acknowledged each
other, maybe even exchanged recipes. Then, together, they
casually jogged toward the dugout to signify the end of the
inning. The ball plopped softly behind them. Nobody, not
even Buck, had ever seen anything like that.

"After a while, losing starts to infect everything," Buck said.
"Pretty soon you can't even do the most basic things in the
world. I've even seen people like that, people in business,
they get into a rut, and they can't get out no matter how hard
they try."

Harvey, the kind of man you only meet at a racetrack, wore
a wrinkled Negro Leagues T-shirt especially for the occasion
of meeting Buck O'Neil. He also wore a crumpled baseball
cap, but that was not in Buck's honor. He always wore that
baseball cap. Harvey looked to be in his seventies, and he
looked vaguely lost, and he introduced himself by saying he
was probably the only person alive who had seen both Joe
Louis–Max Schmeling fights of the 1930s.

"Harvey could not wait to meet you, Buck," Bob Turnauckas
said.

"I have something to show you," Harvey said, and he
reached under his shirt and pulled out a clear plastic bag.
Inside the bag, he had a baseball with all sorts of scribbling
on it. He handed it to Buck. The baseball had autographs.

"I used to stand outside the Polo Grounds and get the

Giants to sign baseballs," Harvey said. "I have three more like this. I think this one—yeah, look. This one has Rogers Hornsby's autograph."

Buck looked hard and, sure enough, there was the signature of Hornsby, one of the greatest players who ever lived. Buck said that Hornsby refused to go to the movies because he thought it would harm his batting eye. Harvey nodded and said that was not the only great player he had met in those days. He met Hall of Famer Carl Hubbell ("Nice guy"). He met Hall of Famer Bill Terry ("Nice guy"). And he met Hornsby. "He was not such a nice guy," Harvey said.

"He could sure hit," Buck said.

Buck asked Harvey if he remembered his first baseball game. Harvey did. He said the game was at Yankee Stadium. He did not remember many of the game details. Instead he remembered how green everything looked. He remembered being carried over the turnstile so his father would not have to pay for a ticket. He remembered standing up on his seat so he could see the field. He also seemed to remember that Babe Ruth hit a home run.

"Isn't that funny?" Buck asked. "People always remember someone famous hitting a home run in their first game. Harvey, I'll bet you and I are the only ones in this whole place who saw Babe Ruth hit a home run."

Harvey nodded, told us again that he saw both Joe Louis–Max Schmeling fights. He then went back to his seat to watch the first race. As he walked away, Bob Turnauckas whispered: "You would never know it, but Harvey is filthy rich. He's absolutely loaded. I don't know what he did. But he lives in an enormous house right around here, right next to Jenny Craig, the diet woman."

Buck placed his first bet on a horse called "Irish Look." He said: "Hey, I'm O'Neil, I'm Irish, let's give it a try." He seemed impressed by the look of Irish Look as he trotted by. Buck was not quite as impressed when Irish Look finished a distant sixth.

"That's all right!" Buck shouted out as Irish Look trotted by again. "You're still my friend!"

Between races, a man sang an off-key version of "When the Turf Meets the Surf," to overwhelming boos. Buck shouted, "You just keep on singing!" He saw a little girl hugging her father, and he said, "That's your daddy, isn't it?" He got into a conversation with a man from Arkansas. His wife, sitting next to him, was from Ohio.

"You married a Yankee?" Buck asked the man.

"I was drunk when I married her," he said.

"You know," Buck said, "I think she was the one who was drunk."

Buck studied the *Daily Racing Form* a bit harder for the second race and found a horse named "Queen of Soul." He said, "Well, that's a winner for sure." Queen of Soul was owned by the diet woman herself, Jenny Craig, which only got Buck more excited. Queen of Soul started fast, but even Buck could tell she would not last. As the Queen died in the stretch, Buck shook his head and looked at Bob Turnauckas.

"Why," he asked, "would anyone bet their own money on this stuff?"

THE NEXT DAY, Buck had an appearance at the Rock Island Art Gallery on Coronado Island. A young artist, Kadir Nelson, was showing a series of Negro Leagues paintings

there. And the owner of the gallery, Mark Cohen, had idol-
ized Buck since he was a child.

"I feel cold," Buck said to Cohen as we rode over the Coro-
nado Bay Bridge.

"Do you want me to turn off the air conditioner?" Cohen
asked.

"No," Buck said. "It's a different kind of cold."

Mark immediately seemed to understand. He told Buck
that the Coronado Bridge, at its highest point, stands two
hundred feet over Coronado Bay, and for many years that
high point has drawn the eye of people who feel lost. More
than two hundred people have jumped off the bridge and
killed themselves. A thousand more have been talked off the
edge at the last instant. "It's one of the biggest suicide spots
in America," Cohen said.

"Well, that explains it," Buck said as he shivered.

Mark then said that the bridge was famous for a dog that
somehow fell off the bridge. He fell into the water, a fireman
fished him out. "He's still alive," Cohen said. "Everybody
knows him."

Buck smiled. "I'll think about that."

Mark Cohen became a writer and the owner of an art gal-
lery, but before that, when he was a boy, he had been a huge
Cubs fan in Chicago. And the most magical day of his
childhood—perhaps the most magical day of his life—had
been when he won a baseball trivia contest on the radio. He
won free tickets to a Cubs game and, even more amazing, the
chance to go on the field before the game began and sit in the
Cubs dugout. He could remember everything about that day.
He remembered how different the Boston ivy on the outfield
walls looked up close. He remembered how smooth the infield

dirt felt beneath his shoes; it was a little like walking on the sand at the beach. More than anything, he remembered how nervous he was as he sat in the dugout next to a tall, handsome coach who asked his name.

"Mark," he said.

"You play ball, Mark?"

"Yes, sir."

"I'll bet you do. I'll bet you are quite the ballplayer."

That, of course, was Buck O'Neil. Now here it was, more than forty years later, and Mark said he felt just as nervous sitting next to Buck in the car.

"I remember I wouldn't stop talking," Mark said. "I'm sure I drove you crazy."

"Kids can't drive me crazy," Buck said. "Impossible."

We arrived at the Rock Island Gallery, and Mark drove us around back. There was an unmarked door there. Mark said, "We'll go through here."

Buck stopped cold. His face went blank.

"Let me tell you something right now," he said. "I don't like going in back doors."

Mark froze. His mouth opened. Horror burned in his eyes. "I'm so ssss—"

Buck put his hand on Mark's shoulder and he laughed and laughed and laughed.

Kadir Nelson waited inside the gallery. He had made a reputation as an artist for the stars. He had illustrated kids' books for Hollywood types like Debbie Allen, Will Smith, and Spike Lee. He worked with Steven Spielberg on the art direction of the movie *Amistad*. He liked doing that. But these Negro Leagues paintings were something, perhaps, a little closer to his heart. He loved sports, especially baseball, and here, he could blend myth and reality.

One of his paintings was of Stuart Jones, a six-foot-six giant everybody called "Slim." Slim Jones was a left-handed pitcher from Baltimore, and he could throw a baseball about as hard as any man alive, even Satchel Paige. They faced each other a few times—Slim and Satchel—and the most famous of their duels happened in 1934, at Yankee Stadium. They were the two best pitchers in the Negro Leagues then, and more than thirty thousand people cheered in the stands. The two men pushed and inspired each other as great rivals can. For the first six innings, nobody reached base against Slim Jones. He would give up three hits in all, Satch gave up six, twenty-one men had struck out. Witnesses would say that no two men had ever thrown baseballs any faster. Darkness brought the game to a close with the score still tied.

Not long after that game, though, Slim Jones's fastball began to slow. He drank a lot. He palled around with shady characters. He constantly asked for money. The Negro Leagues historians say he did not win a single baseball game in 1935. And he despaired. Three years after that, he sold his winter coat for his last bottle of whiskey and he was found on the street, frozen and dead. Kadir Nelson's painting, called *Low and Away,* captures him during his great game against Satchel Paige, and Slim Jones is outsized, he towers over the scene like a New York skyscraper, but the painting also has a certain sadness.

"I like that one," Buck said as he looked at Slim Jones. "But this is my favorite."

He pointed to a painting called *Coooool Papa Bell.* In it, an enormous Cool Papa Bell slides into third base. He is stretched out, elongated, so that he must be fifteen feet tall. Cool Papa's leg slides into the bag just as the baseball is about to land in the third baseman's glove.

"He's safe," Buck said as he pointed at the painting.

"You got the point right away," Kadir said.

Buck sat in a metal folding chair, and people surrounded him and asked baseball questions. Then a familiar thing happened. The questions stopped. And people started telling Buck their own stories about baseball. This never failed. There was something about Buck that seemed to invite people to tell him their stories.

The television reporter who had come to interview Buck stopped in the middle of the interview, put down his camera, and told Buck how the first game he ever attended was rained out. "I cried like a baby," he said. "But I still have the program."

Another guy told Buck a story about the 1971 All-Star Game. In that game, Reggie Jackson hit the ball onto the roof of old Tiger Stadium. The ball had clanked off one of the light towers on top of the stadium. A Baseball Hall of Famer, Al Kaline, said he had never heard a sound quite like the one made when the ball hit Reggie's bat.

"I was watching that game," the guy said, "and I put a tape recorder next to the television so I could record the announcer's call. Then Reggie hit the ball . . ." and the man clutched his heart and said that in the days and weeks afterward, he would play the tape again and again and again, until he knew every crackle and buzz of the tape. He listened over and over to the sound of Reggie Jackson hitting a baseball out of Tiger Stadium and beyond imagination.

"You know what play I love?" Buck said. "The triple. That's my favorite play. Someone hits a home run, what happens? Everyone stands around and watches the ball go. Then the guy jogs around the bases. But, man, someone hits a triple . . ."

And then Buck spoke a haiku:

> *Everyone's running.*
> *The whole field bursts to life, man,*
> *Best play in baseball.*

Outside, a cool wind blew off Coronado Bay. It was still summer by the calendar, but fall chilled the air. On one side of Buck, Mark Cohen still talked about how nervous he felt when he was a boy and was trying to get autographs from the Chicago Cubs. On the other side of Buck, Kadir Nelson talked about how scared and amazed he was when he met the all-time home-run leader Hank Aaron. A young couple on their honeymoon walked into the gallery, three women who had just finished their book club lunch walked in, an older couple who had just been outside arguing about his driving walked in, and soon they all gathered around Buck O'Neil, and they either talked about baseball and being young or they listened.

"What did I tell you," Buck said as the gallery burst into life around him. "People say baseball's dead. Baseball doesn't die. People die. Baseball lives on."

AUTUMN

I GOT A RIGHT TO SING THE BLUES

Monte Irvin and Buck O'Neil sat on a bench in front of the stadium in San Diego and watched the baseball fans walk by. The people were headed to the evening ballgame, the San Diego Padres against the Philadelphia Phillies. The baseball fans wore caps and authentic Padres jerseys. Many of the jerseys featured the name "Greene."

"Who is Greene?" Monte Irvin had asked.

Buck shrugged. Khalil Greene was a blond-haired Padres shortstop who, even under his baseball cap, had the unmistakable look of a surfer. He fit San Diego. The fans, so many in their Khalil Greene jerseys, walked slowly. There's no rush in California. Some of the fans drank out of beer cans. Some carried their baseball mitts.

"Nobody uses the word 'mitt' anymore, have you noticed that?" Buck asked.

"Well, they still say 'catcher's mitt,' " Monte Irvin said.

"Yes they do. But for everything else, they say 'glove.' They don't say 'mitt.' We used to call all the gloves 'mitts.' "

"We did? I don't remember that. I think we called them 'gloves' sometimes."

"I like the word 'mitt,' " Buck said.

Nobody seemed to notice Monte and Buck, which suited them fine. This gave them a chance to talk about nothing. They had come to be part of the San Diego Padres' Annual Salute to the Negro Leagues, and they both knew that in only a few minutes they would be rushed off to make appearances and do interviews and sign autographs, and there would be no chance to talk.

"I see we lost Double Duty," Monte Irvin said.

"Yeah."

"You going to the funeral?"

"Yeah. You?"

"No. I can't make it. Duty lived a good life."

"What you talking about? He lived one hundred and three years. He lived a great life."

"Never stopped living."

"No sir. You're right, Monte. Never stopped living."

Monte Irvin leaned on his cane. He was eighty-five, and only recently his body had started to betray him. Doctors kept telling him he would not get better unless they operated, and Irvin believed them. He had decided he would not get better.

"You ever stretch, Buck?"

"Stretch?" Buck said. "What do you mean?"

"Can you stretch your back?"

"Well, yes, I can still stretch. But once I start stretching, I can't get back up."

"I know what you mean. I can't stretch at all. Sometimes that's all I want in the world. Just a good stretch. It's hard growing old."

"Better than the alternative, Monte."

"I suppose it is that."

Baseball fans kept walking by these two men who had done so much in the game. A few months earlier, Monte Irvin—a baseball Hall of Famer, a man who had once played ball like Willie Mays—said he sometimes felt bad about things. He said: "Sometimes I think, *Nobody saw me when I could really play.* I guess that's human nature." Now, Monte sat on a park bench and breathed in the ocean air and talked easily in his invisibility. He seemed quite happy that people wanted to see Khalil Greene.

Monte Irvin was probably the only player to be a star in both the Negro Leagues and the Major Leagues. Willie Mays and Hank Aaron were just kids when they played for their brief times in the Negro Leagues. Satchel Paige was an old man when he played in the Major Leagues, and even though he was occasionally brilliant, he mostly served as a publicity stunt.

Monte Irvin was thirty-two years old when he got to play his first full season in the big leagues. He had already lived a baseball life. He played remarkable baseball in the Negro Leagues. He won a baseball championship in Cuba. He was a hero in Mexico and something bigger than life in Puerto Rico. He had also gone to fight in World War II. After turning thirty, he signed with the New York Giants and went to play in the Minor Leagues. He could feel time running out, though, and in 1950 he told the Giants he was ready to play in the Major Leagues. The Giants management told him they would decide when he was ready, and they shipped him back to the Minor League team Jersey City. In eighteen games in Jersey City, Monte Irvin hit .510 with ten home runs. The Giants decided he was ready then.

He was ready, but he was not the same. The war and the years had taken something out of him. In 1951, he showed just a little bit of the player he had been as a young man. He led the National League in runs batted in. He hit eleven triples and stole twelve bases on tired legs. He also hit twenty-four home runs, and he finished fifth in the league in hitting. Irvin helped the Giants catch the Dodgers in the last days of the pennant race, the greatest comeback in baseball history. In the World Series he hit .458 against the mighty Yankees. He stole home. He was, in short, one of the best players in the Major Leagues. Even in that remarkable year, though, Irvin could not help but feel that something had been taken from him.

"I was a different player by then," he told Buck as they sat in the San Diego dugout while the Philadelphia Phillies players took batting practice. "I was still good. But I was not the same player."

"I know," Buck said.

"When I was young, I played a big center field," he said. "Like Willie Mays."

"I remember," Buck said.

Irvin had been an amazing athlete at East Orange High in New Jersey. He was a four-sport star. He was an unstoppable basketball scorer and he set the state record in throwing the javelin. In baseball he felt limitless. Nobody could hit a fly ball Monte Irvin could not run down. Nobody could throw a pitch past him. He felt invincible until the war. He spent three years in the army. When he came home from Europe, his body did not feel quite the same. He could still play brilliant baseball—he could still hit home runs, still steal bases, still catch most of the fly balls hit out there—but it was not effortless, not like it had been before.

"I'm not complaining," Irvin said. "I mean, I lived a good life. Better than most guys in the Negro Leagues. I got to play in the Major Leagues. I got to play in the World Series. I'm not complaining. It's just that people used to tell me how good I was, and I would tell them, 'You should have seen me when I could really play.'"

"I saw you, Monte," Buck said.

"And?"

"You could really play."

"That's all I was saying," Monte said, and he smiled too.

KENNY LOFTON, THE Philadelphia center fielder, walked over to hug Buck. He said, "We've got the new Josh Gibson. I'm not kidding with you, Buck. I'm going to bring him over. I want you to meet the new Josh Gibson."

"Bring him on over," Buck said.

Lofton ran off and came back a few seconds later with a huge hulk of a man named Ryan Howard. His baseball card listed him at six foot four and 252 pounds, but in the bright sunshine he looked about twice that big. Howard had a sheepish look on his young face, as if he felt bad for taking up so much space on this earth. Lofton said Howard's face transformed when he was hitting. The sheepishness turned cold, and though Ryan Howard had only been in the Major Leagues for a few weeks, pitchers feared him. Lofton could not help but be giddy about him.

"I've told Ryan that he's like a left-handed Josh Gibson," Kenny said. "You should see him swing the bat, Buck."

"You got some power, young man?" Buck asked.

"A little bit, sir," Ryan Howard said.

Buck shook his head. "Don't be ashamed of your power," he said. And then, without knowing it, Buck offered up another haiku.

> *If you got power,*
> *Don't hide it for nobody.*
> *Swing the bat hard, son.*

Kenny Lofton said, "That's what I've been trying to tell him. But it sounds better when you say it, Buck."

In the next year, Ryan Howard would hit a 500-foot home run in Philadelphia. He would win the Home Run Derby at the All-Star Game. He would smash the Philadelphia Phillies record for most home runs in a season. He swung the bat hard.

People wandered up to Monte Irvin for autographs and photographs and he sat on the railing by the dugout. He obliged graciously. But every time they walked away, he returned to the railing and tried not to catch anyone's eye.

"Didn't they try to sign you before Jackie Robinson?" a man asked Irvin.

"The Giants did talk about signing me in 1945," Irvin said quietly enough the man had to lean forward to hear. "It was right around the same time as Jackie. But I wasn't ready then. I had just gotten back from the war. I could have been first, I think. I could have handled it. But it wasn't meant to be. It was Jackie's time."

Buck and Monte headed upstairs to do a television show together. Buck walked across the field. He stopped to sign a few autographs. He shook hands with some people in the stands. He caught a foul ball on one bounce and tossed it up to a young fan. Monte Irvin leaned hard on his cane and tried to walk across. He made it five steps and stopped. A truck picked him up and drove him through the outfield.

"Hard thing, growing old," he said again as the truck passed Buck O'Neil.

"I remember those rickety old buses," Monte Irvin said when the television hosts asked what he remembered best about the Negro Leagues. Monte talked about the bus rides, and how they would shake up his insides. He said that the players would wash their clothes in the morning and dry them by holding the clothes out the window on the ride to the next town.

Buck grimaced. He never did like hearing stories that made the Negro Leagues sound second-class. The longer we were together, the more troubled he seemed by those stories. He supposed people already thought the Negro Leagues was second-rate baseball, and these stories about wobbly buses and hobo ways strengthened those feelings. One of the radio hosts asked Buck to tell the story about how, when he was very young, his team would have so many people in one car that some had to ride on the bumper. Buck shook his head.

"No, no, that was not the Negro Leagues," Buck said. "I did that, but that was in our younger days, when we were barnstorming, when we were playing on our own semipro teams. The Negro Leagues, we had buses. They were some of the best buses money could buy."

Monte Irvin looked over at Buck.

"You must have had better buses than us, Buck," Monte said.

"We probably did," Buck said.

"Anyway," Monte said, "Buck has a better memory than I do."

Monte Irvin talked about timing. He said that had he been born a little later, he would have spent his whole career in the Major Leagues, and people might think of him as one of the greatest players who ever lived. They might think of him the way they think of Willie Mays, Joe DiMaggio, and Mickey Mantle. Then again, he said, had he been born a few years earlier, he might not have been known at all. He would have spent all his life in the Negro Leagues, playing baseball on rock diamonds in small towns. The only way anyone would know him would be through whispers and myth.

"Either way, though, I would have gotten to play ball, and that's all I ever wanted to do," he said. "I don't feel sorry for myself. I got to play."

Nearby, Buck looked out on the field. He did not seem to be listening.

"You know what I always wondered," Monte said. "Why did they think we couldn't play? The ball was the same size. The bats weighed the same. The fields were no smaller. The fences were no lower. It was still sixty feet six inches to home plate. Still ninety feet to first base. Why did they think we were inferior? Why did they think we could not play this game?"

Buck never turned away from the field. He whispered:

It makes no sense,
Hate.
It's just fear.
All it is.

Fear something different.
Something's
Gonna get taken from you,
Stolen from you,
Find yourself lost.

Monte Irvin them remembered a story. When he came out of high school, the Major League scouts knew about him. Even in those days, the good scouts felt a curiosity about black baseball players. Buck liked to say a scout is drawn to talent the way a comedian is drawn to the stage. He is drawn to small towns and overgrown fields and back roads. The times were different. And scouts were no more or less racist than anyone else. But the good ones couldn't stay away.

Scouts watched Monte Irvin play baseball. He knew they were there, or at least he sensed it. No baseball scout ever talked to him, of course. But he noticed a few white men in the stands, watching him closely, writing down things—reports they never intended to file—and then, when the game ended, they slipped away quietly. Monte Irvin knew they were there. He just never thought much about it.

Many years later, he played for the New York Giants. Years after that, he was inducted into the Baseball Hall of Fame. He worked for Major League Baseball. He became a member of the Hall of Fame veterans committee, and worked along with Buck O'Neil to get Negro Leagues players into the Hall of Fame. Monte Irvin became one of baseball's most respected men.

And in those later years, he ran into Horace Stoneham, who had been the owner when Irvin played for the Giants.

Stoneham was, by most accounts, a gentle man who had, by circumstance, become despised by Giants fans on both sides of America. He had moved the Giants from New York to San Francisco, leaving behind angry fans who cursed his name. In San Francisco, he had traded an aging and tired Willie Mays at the end of his career, infuriating West Coast fans. Stoneham was in his mid-eighties when he saw Monte Irvin. He would die soon.

"You know we sent a scout out to see you in high school," he told Irvin.

"Yeah, I thought you probably did," Irvin said.

Stoneham then told a story. It turned out that Irvin's high school teachers had called Stoneham personally and said: "We've got a player here you would not believe." Stoneham did not intend to integrate baseball then. It was 1938—more than fifteen years before Rosa Parks refused to go to the back of the bus—and he knew as well as anyone that baseball had no stomach for integration. Stoneham was no pioneer. But he did feel curious. He had to know about this kid. He sent out a scout, and then another, and then another to see Monte Irvin play.

"What did they tell you?" Irvin asked Stoneham.

"They told me you were the next DiMaggio," Stoneham said.

Irvin had spoken about this conversation many times before. He did not get much joy out of telling it. Words are only words. Yes, he played like DiMaggio in those dark years before Jackie, but people did not see him. There are no surviving films. The *Baseball Encyclopedia* does not show his statistics from 1941, when he played his best baseball. And it never will.

"He told me it was too soon," Irvin said. He shrugged. He

said Stoneham felt regret. The old man said he wished he had shown more courage as a young man.

"Those scouts told Stoneham I could have been one of the best ever," Irvin said.

"You *are* one of the best ever," Buck O'Neil said.

A FUNERAL IN CHICAGO

Buck O'Neil showed up at the airport early as always. He wore a dark blue suit with matching blue shoes. He announced while boarding the plane that this would not be a sad day, no sir. He would not allow any sadness on this day. Early-morning sunlight poured through a small round window and hit Buck's eyes. The sun had just come out, but the weather was warm and the plane was hot. Passengers wrestled with the air-conditioner vents above their seats. They could not get cool air to blow no matter how much they twisted.

"Double Duty's funeral," Buck said softly as the plane took off. "Yes sir, this will be a good day."

The writer Damon Runyon gave Ted Radcliffe the name "Double Duty." Runyon was a sportswriter then, and he watched Radcliffe catch one day and pitch the next. That happened during the 1932 Negro Leagues World Series. Runyon said "Double Duty" was worth two admissions. Radcliffe loved that nickname. He carried it with him for more than seventy years. It was, in many ways, his most prized possession.

Double Duty used to say he was the greatest player who ever lived. In his younger days he said it with a sly grin, as if he expected everyone was in on the joke. In his last years, the grin was more or less gone. As a catcher, he once threw out Ty Cobb trying to steal. This infuriated Cobb, who was well known for being a racist. Cobb glared hard at the catcher, and only then did he see that, across Duty's chest protector, Duty had scribbled the words "Thou Shalt Not Steal." As a pitcher, Duty threw every illegal pitch imaginable: he spit on the ball, cut it, rubbed shoe polish on it—anything to get people out. He was, by most accounts, effective. Most Negro Leaguers remembered him being a fair hitter. Double Duty, of course, remembered being great.

As a talker, though, everyone agreed: Double Duty was king. He played for more than thirty teams—Duty never had much faith in contracts—and he managed to irritate and amuse just about every player who ever put on a Negro Leagues uniform. All game long, Duty jabbered and insulted and encouraged, he told stories and jokes, he bragged to the point where other players longed to hit him in the head with a bat. No one who played baseball with Double Duty Radcliffe forgot him.

Double Duty disappeared for many years after his playing days. He and his wife scraped by on the South Side of Chicago until 1990, when they were attacked and robbed in their small home. The story made the news. Double Duty was rediscovered. He began to appear at games and functions, and he told stories, he worked on a book, he signed autographs. Double Duty could talk. The year he turned one hundred, Duty threw out the first pitch at a Chicago White Sox game. Two weeks before he died, he threw out the first pitch before a game at Rickwood Field in Birmingham, the oldest ballpark in America. The cancer had seared through him by then.

"There's nothing sad about a man living one hundred and three years," Buck said as the plane bounced in the hot air above Chicago. "And Double Duty lived, no doubt about that."

The car skimmed over the potholes and raced through yellow lights. There was no hurry, but this was the South Side of Chicago, and the driver figured there was no reason to linger. Pawnshops. Liquor stores. Check cashing. Out the window, two men staggered along as if they were tightrope-walking on the sidewalk. Broken glass. A woman sat on a blanket in front of a porch. Buck watched close.

Buck talked about his old friend Willie Spooner, who ran a series of camps for young baseball players in Baton Rouge. Buck traveled down every year, and so would his players—Ernie Banks, Lou Brock, Billy Williams, George Altman, and the rest. The camp was supposed to teach kids about life through baseball.

> *Nothing better*
> *Than baseball. For kids.*
> *Teach them all the lessons.*
> *How to be a teammate.*
> *How to be a man.*
> *Nobody does it for you.*
> *Gotta stand up.*
> *I remember Willie*
> *Used to tell those kids in Baton Rouge*
> *It's better to steal second*
> *Than to steal an apple.*

We arrived early at the Apostolic Church of God on Sixty-third and Dorchester. A woman stood outside, a woman Buck did not know, but she shrieked "Buck!" and rushed over with tears in her eyes. She gave Buck a big hug. Buck, for once, did not seem in the hugging mood.

"It's so sad," she said between sobs.

"It's not that sad," Buck said. "The man lived a good life."

Al Spearman sat in a pew in the Apostolic Church of God when Buck arrived. He clutched several manila folders. Spearman played in the Negro Leagues for a short while before getting lost in the Minor Leagues in the 1950s. He had been a good player, even if he did not reach the Major Leagues. Lately, though, Spearman had not talked much about his own career. He had grown obsessed with a gentle-looking, white-haired man named Johnny Washington who was going around Chicago telling everyone about his own days as a Negro Leagues player. Al Spearman had grown convinced this Johnny Washington was a fraud. Buck had heard the charge many times.

"There was a Johnny Washington," Al Spearman said, plunging right in. "I remember him, played in the 1930s and '40s, but this isn't him, Buck. It isn't him. This Johnny Washington doesn't exist." His eyes pleaded. His voice grew louder. People turned around.

"This is not the time, Al," Buck said softly.

"Look here," Spearman said, and he pulled out a photograph—it was of Al, Buck, and Double Duty standing at Comiskey Park in Chicago before a game. Johnny Washington stood there too. Unlike the others, he was not wearing the red credential needed to get on the field.

"You see!" Spearman shouted. "He's not wearing the red tag! He hopped the fence! Look at this photograph. This is proof, Buck. Living proof. This man is a fraud. He didn't play. He's going around telling people all the suffering he went through. . . . He didn't play, Buck. He didn't play."

Buck put his hand on Al's shoulder and looked toward the photograph, but he did not focus on it. Buck appeared to be looking through the picture at the distant wall. Buck looked toward the photograph only to avoid seeing the tears in Al Spearman's eyes.

A CERTAIN KIND of man will try to take control at a funeral, especially at the funeral of famous men. This was one of the reasons Buck stopped going to funerals. People he did not know pushed him and pulled him, took him over to the cameras, introduced him to famous players as if they had never met before. He did not like being handled. This day was no different. A man Buck did not know walked over with Billy Williams. The man said: "Buck, I don't know if you've ever had the pleasure of meeting Billy Williams." Billy and Buck smiled at each other.

"Yeah, we've met," Buck said.

"This man saved my career," Billy said.

"Oh," the man said, and he did not look the slightest bit embarrassed. The man stepped away and walked over a few minutes later with another Chicago baseball legend, Minnie Minoso. "Buck, I don't know if you've had the chance—"

"We've met," Buck said with slight annoyance in his voice now. "How are you, Minnie?"

"Good, Buck, good."

"We need to go to Cuba together again, Minnie."

"We will. It will open up, Buck. We will go."

This went on for a while. The man wandered the church, ordering people around, introducing people who had already met, promising the reporters and cameramen that he would bring the players to them, and acting as if he were running things, though nobody seemed to know exactly who he was or what connection he had to Double Duty. At the same time, another man walked quietly through the church and everyone knew him. He was a character named Cleve Walker. Cleve was one of Double Duty's best friends. Being a friend seemed to be Cleve Walker's occupation. He tried baseball years ago, though nothing much came of it. Cleve first met Buck when he tried out for the Chicago Cubs. He never did play much ball, but he came to know the ballplayers, and then he knew other celebrities. He traveled with Muhammad Ali. He knew most of Chicago's biggest men. He seemed to be in the background of many photographs.

"Nobody knows exactly what Cleve does," Buck said in wonderment, "but he is beloved by everybody." As if on cue, Cleve walked over to Buck, they embraced, and it became clear why everyone liked him so much. Cleve listened. He looked interested. While other men around him told their own stories and glanced around the room to see if there was someone more famous nearby, Cleve laughed and nodded and acted as if no one else mattered. He never interrupted. He waited for silences.

"I suppose I was just about the last one to see Duty alive," he said in one of those silences.

Everyone stopped and looked at Cleve.

"Yeah, he was not doing well. All those years, we always did the same thing. Duty would come up to me and say, 'I'm taking you deep today, Cleve.' You know, like he was going to hit a home run off me. And I would always say, 'No way, Duty, not today.'

"So I go up to him on his last day, and I'm not even sure he knew who I was. He was in bed. He looked bad, you know. It looked like the end. And I said to him, 'No way you're taking me deep today, Duty,' and I pretended to pitch. He tried to say something, you know, but he couldn't talk. So then, I see his arm go like this—"

Cleve pushed his arm out, the same motion Ralph Kramden used to make when he talked about sending Alice to the moon on *The Honeymooners*.

"And Duty mouthed the words *I'm taking you deep*. That was the last time I saw him."

JOHNNY WASHINGTON SHOWED up. If a casting director needed someone to play the role of a former Negro Leagues player, Johnny Washington would have fit nicely. He looked distinguished, with a crisp black suit contrasting with his white hair. He walked up and down the aisles, and he shook hands with everybody. He introduced himself as "Johnny Washington, former Negro Leagues player" to those few people who did not know him. Al Spearman walked three or four steps behind him. Spearman looked like a man who had lost faith in the world.

"Hi, Al," Johnny Washington said with a smile.

"I'm not talking to you," Al Spearman said.

"Yes, this man's going around telling people that I didn't play in the Negro Leagues," Washington said to those surrounding him. He said this in an untroubled voice, as if he believed his accuser was clinically insane but quite harmless. This set off Spearman to a new level of rage.

"You didn't play!" Spearman shouted. "You didn't! You're a fraud! You're a disgrace to all of us who really did play!"

Washington smiled again. He showed no anger at all. His eyes were filled with wonderment—as if this man were shouting in another language. Buck watched all this for a second and then stood up and took Spearman's arm. "This is not the place, Al," Buck said. The reverend walked behind Double Duty's casket. There was a red fedora on top. The funeral was about to begin.

A FRIEND OF Double Duty's played "Amazing Grace" on the trombone. He remembered that Duty used to say to him, "Get Jesus and play that jazz, you'll be all right."

Sean Gibson then stepped on the stage, and Buck's eyes opened wide. Sean was the grandson of Josh Gibson, probably the greatest slugger ever to play in the Negro Leagues. Sean explained that Double Duty would tell him things because he looked so much like his grandfather. "He thought I was an angel," Sean said, and four rows back, Buck O'Neil nodded. "He does look just like Josh," Buck whispered. "It's scary. He looks more like Josh than Josh did."

"Duty told me so many stories," Sean said. "Some of those stories, though, I can't tell in church."

Hymns were sung. Acknowledgments were read. The governor of Illinois had sent his condolences and so did the owner of the Billy Goat Tavern in downtown Chicago. Kyle McNary, a middle-aged man from South Dakota, spoke a few words about Double Duty. Kyle worked in construction, but his passion was the Negro Leagues. When he met Double Duty and listened to just a few stories, Kyle fell in love. He wrote a book about Double Duty. "He was my hero," Kyle said.

The reverend spoke again. "None of us are here because Double Duty died," he said. "We're here because he lived."

Buck was called to the stage. Nobody had told Buck that he would be asked to speak, but he had expected it. He had practiced in the car on the way over. He had said:

> *This is not*
> *A sad occasion.*
> *No sorrow when a man*
> *Lives a full life.*
> *Don't cry.*
> *Save your sadness*
> *When they're taken young,*
> *Before their time.*
> *Man lives to one hundred and three,*
> *Rejoice.*
> *Got to do everything,*
> *Feel everything*
> *There is to feel in this world.*
> *Don't want to live forever.*

Buck said some of those things at the microphone. More, though, he talked about what it meant to play baseball in the Negro Leagues. It was about changing the world. An organ played, the tones climbing and cascading in rhythm with his words. There were flowers on the casket and blue and red sunlight poured through a stained-glass window. Buck began to preach.

"When Jackie Robinson crossed the color line, that was the beginning of the modern-day civil rights movement," Buck said. "I say 'modern-day civil rights movement' because the first civil rights movement began down in Egyptland, didn't it?"

A woman yelled "Amen!" A man shouted "Preach it, brother!"

"My God needed a man. A strong man. And my God sent a man named Moses. That's what my God did." ("Amen!" "You preach, Buck!" "Hallelujah!") "Then, here in this land, we were in bondage. Weren't we?" ("Yes we were!" "Amen!") "My God needed a strong man. My God sent Abraham Lincoln." ("Yes he did!") "But ol' Abe Lincoln couldn't do it alone. So my God sent Frederick Douglass. My God sent . . . Sojourner Truth. Harriet Tubman. Yes. That's what my God did."

The sound of the organ rose and fell. Amens echoed.

"And, in my time, we were in bondage. We could not eat in the white restaurants. We could not sleep in the white hotels. We could not play baseball because of our beautiful tans. God needed a man. He sent us Jackie Robinson."

Buck's voice grew quiet.

"But before Jackie Robinson, there were men who played baseball. And we were good. We could play, man. Double Duty Radcliffe could play. People who saw us, man, we could play. We made a difference in this world. Duty made a difference."

And then he sang his song, his verse, "The greatest thing in all my life is loving you." At the end, Buck was crying.

ON THE WAY to the small reception in the back room, someone asked Buck O'Neil for an autograph. He said, "I don't think you should do this here." He signed it anyway.

"God bless you, Buck," a woman said as she hugged him.

"He's already done that," Buck said.

Buck took some fried chicken and sat down. He ate slowly and with his head down. Commotion rained around him. Women cried, men laughed, strangers hugged. Al Spearman walked around, looking for a table, and then a young boy, ten years old, walked up to him. The boy asked: "Are you somebody?"

"Excuse me?" Spearman asked sharply. The boy froze.

"I mean are you somebody famous?" the boy asked. But Spearman did not hear. He saw something in the boy's hand. "What's that?" Spearman asked. And the boy showed Spearman a baseball card. Spearman's eyes bulged. His face blushed. He said, "Let me have that for a minute, boy," and he swiped the card, and he started walking fast toward Buck. It was a card of Johnny Washington.

"Look at this! Look at this!" he shouted. People blocked his path. Spearman could not get around the table to Buck, so he was now shouting at anyone who would listen. He shouted about how this man, this fraud, was handing out his baseball cards like he was somebody, like he had played on those fields with rocks scattered along the base paths, like he had slept on buses bumping along two-lane roads, like he had eaten stale sandwiches in towns where restaurant owners would not serve black men. Spearman looked as if he might burst out of his skin when a man—the man who had acted as if he were in charge—walked over.

"Al, you are going to have to quiet down," he said.

"But this Johnny Washington—"

"Al, quiet down. This is not the place. We've all heard you already."

"But, no, this man is—"

"Al, you will have to quiet down or I'm going to have to make you quiet down."

"What, you're threatening me now? You're threatening me?"

Spearman walked away shouting, "This man is threatening me! He's threatening me for telling the truth! Said he's going to 'make' me quiet down! This man here"—and he pointed, and shouted. Buck quietly ate his chicken. Al Spearman thundered until no one listened anymore.

A MAN WALKED over to Buck O'Neil and said, "I don't mean to interrupt you, sir, but you are an inspiration to me."

Buck looked up and said: "Do you remember the first baseball game your father ever took you to?"

The man looked surprised. He said he did remember. It was 1951 at Comiskey Park. He was five years old. The man did not remember who the White Sox played, who won the game, or even whether it was day or night. He thought it was a day game, a Sunday, though what he really remembered was that Minnie Minoso hit a home run. Buck had been right. Someone famous always hits a home run at a boy's first baseball game.

"You know, Minnie Minoso is here," Buck said.

"I did know that. I've met him a few times. He is my hero."

"Mine too," Buck said. "Great hitter. Great player."

"He should be in the Hall of Fame. And so should you."

"Isn't it funny? Everybody remembers going to their first baseball game with their father. They might not remember going to their first day of school with their mother. They don't remember their first football game or their first Thanksgiving dinner. But they always remember going to the baseball game with their father."

"Why do you think that is, Buck?"

"I don't know," Buck said. I had never heard him say those three words before.

It was time to go. The car was on its way to take us back to the airport. Buck said: "Wait one minute. There's something I have to do."

AL SPEARMAN WAS alone and ranting in the hallway when Buck walked over. He put his arm around Al's shoulder and said, very softly, "Al, I want you to listen to me."

"Buck, I—"

"Al, let me say something. You know he didn't play. We know the truth. It's all right. It's all right. It's all right. It's all right. . . ."

He kept repeating the phrase, over and over—it's all right—as if he wanted to hypnotize Al Spearman. He kept trying to get Al to look into his eyes, though Al could not seem to focus for very long.

"It's all right, Al," Buck said. "Maybe this man needs this."

"Needs this?"

"Maybe this man needs people to believe that he played in the Negro Leagues. Maybe this is what is keeping this man alive, see? It's all right. Let the man be. We know."

"But, Buck, this guy is a—"

"We know. It's all right. You know. I know. That's all that matters. It's all right."

For a moment, it seemed like Al Spearman's demons had passed. He slumped his shoulders and seemed to nod. And then, suddenly, he began shouting again, "No, it's not all right! Listen here, this man, he's a fraud, I have pictures, look at this picture." And he pulled out the same photograph, the one that showed Johnny Washington without a red credential. Buck shook his head and walked away. Al followed Buck into the bathroom. Spearman's voice got louder and louder, and you could hear him through the door. He came out still shouting.

Buck turned to Al Spearman and said, "I've got to go now."

"You'll think about what I told you. This man, he's tainting all of us. He's a fraud."

"I'll think about it, Al." And Buck pushed open the glass doors in the front of the church. He turned back to Spearman and whispered—so softly that Spearman could not have heard—"Let go, Al. Just let go."

THE RIDE BACK to the airport was quiet. The sun burned hot, the car air conditioner blasted cold air through its vents, and Buck looked out the window at the liquor stores and abandoned lots and weeds and places selling gyros. A man and a woman walked down the street, hand in hand. Each of them clutched a brown paper bag wrapped around a bottle.

Buck said: "It's sad to see how many people live in pain."

WASHINGTON

Buck O'Neil could not stop watching when the baseball steroid hearings raged on television. He felt strangely connected to it all, though few people knew less about performance-enhancing drugs. He read stories about the drugs—human growth hormone, the cream, the clear, andro, creatine—and it seemed like Chinese to him. He never could understand how much any of these substances helped a man hit a baseball or pitch one. He did not want to know. People were always surprised that Buck did not have strong feelings about how bad steroids were for baseball. He did worry about the kids ruining their bodies, but the cheating part did not move him much. In the Negro Leagues, he had known players to bend the rules to win—they corked bats, spit on the ball, popped amphetamines, stole signals, and even loaded up on coffee for the caffeine. They wanted to win. "The only reason players in my time didn't use steroids," he would say sometimes, "is because we didn't have them."

Something else about the steroids hearings chilled him. Great baseball players—men he knew—wore dark suits. Lawyers flanked them. Members of Congress asked questions. The players offered statements about how much they loved baseball. And Buck felt an overpowering sadness—it was like base-ball was on trial. There was Rafael Palmeiro, the great first baseman who in the very next season would become only the fourth man ever to collect three thousand hits and five hun-dred home runs. Palmeiro talked about coming over with his family from Cuba when he was just a baby. He said his family fled the terror of Castro so he could live the American dream.

"I have never used steroids, period," Palmeiro had said, and then, punctuating his statement Bill Clinton style, he pointed at the camera.

"He's lying," Buck said.

Sammy Sosa spoke. He was an American phenomenon just a few years before. Sosa had come from the Dominican Re-public and he shared his quintessential baseball tale. He had shined shoes to help his mother and six brothers and sisters survive. Sosa started playing baseball at fourteen—he had boxed when he was younger—and he used a milk carton for a glove and a pipe for a bat. Sosa took to baseball. In 1998, he and Mark McGwire hit home runs at a pace previously un-known in the Major Leagues. They both broke Roger Maris's hallowed record of sixty-one home runs, and they just kept on hitting baseballs over fences. Fans were captivated. Stadi-ums were packed. Baseball, after almost falling into the abyss over a labor problem, was the national pastime again. Sosa rekindled so many of the old feelings in Chicago. He was like Ernie Banks. He played the game with unadulterated joy. He ran around the outfield before every game. He tapped his heart and blew kisses to heaven after each home run. When talking about the game, he said: "Baseball has been berry, berry good to me." America was charmed. Sosa hit sixty-six home runs that year, which would have been a record had it not been for McGwire. The next year, he hit sixty-three more home runs, and two years later, sixty-four more. Few wanted to delve too deeply into the mystery of how a skinny kid who had never hit more than forty home runs in a season could suddenly look like a professional wrestler and slug home runs at Babe Ruth's pace. As Buck would say, "We wanted to believe. We needed to believe."

"I do a lot of charity work with young people," Sosa told the congressional committee. Sosa said he would never put anything dangerous like steroids in his body.

"He might be lying too," Buck said.

Then there was the star witness. Mark McGwire. While Sosa became America's mascot—rags to riches and all that— McGwire became Superman. He stood six foot five, looked sturdier than the Chrysler Building, and before games, during batting practice, he mashed monstrous home runs that soared and crashed against the cheapest upper-deck seats. Nobody had ever hit a baseball like this. Thousands of people left work early, raced home to grab their children, and rushed to the ballpark hours early just to watch McGwire take batting practice. McGwire became living history. Seeing him hit homers was like watching Neil Armstrong step on the moon. "You will remember this all your life," fathers told sons.

There was, even then, a nagging feeling that McGwire had pushed beyond the limits of natural power. In his record-breaking year, a reporter noticed a bottle of androstenedione in McGwire's locker. Andro, as it would be called to help the newspaper headline writers, was a legal substance then in society and baseball. Other sports had banned it. Doctors quoted in the papers said it was a "natural hormone" and a "precursor to testosterone." Weight lifters nodded knowingly—they knew andro worked pretty much like a steroid. But in that magical season, nobody wanted to hear about the dungeons and dragons of Mark McGwire. He hit the record-breaking home run and cameras flashed, and he lifted his son at home plate. He hugged the family of Roger Maris. He hit home run number seventy on the last day of the season, and he ran the bases to deafening cheers. A portion of Interstate 70 in Missouri was named for him. The reporter who found the andro was scolded for snooping in other people's lockers.

In the hearing room barely more than six years later, McGwire looked thinner. It wasn't just his body. He looked hollowed somehow.

"I live a quiet life with my wife and children," McGwire said. His voice was pleading.

McGwire offered to be a spokesman against steroids. He offered to help teach children that using steroids was wrong. But he did not admit using steroids. He did not explain why using steroids was wrong. He did not say much. There was something vaguely admirable about McGwire—unlike the others, he did not proclaim his innocence—but the longer his testimony continued, the longer he refused to say anything, the more absurd he became, until finally he crossed beyond absurdity and became pathetic and sad.

"What I will not do, however, is participate in naming names," McGwire said.

"I'm not here to talk about the past," McGwire said.

When he stepped down after his testimony, Mark McGwire was no longer Superman. No one except for the truest believers had any doubts left. He had used steroids. They all had. Palmeiro, just weeks after venting and pointing at the camera, flunked a Major League steroid test. For days afterward, weeks even, the news stations and newspapers focused on this "black eye for baseball." Buck stared at the television even after the hearings were over. He realized why he felt so sad. Nobody had spoken for baseball.

MONTHS LATER, BUCK O'Neil was sitting in a black car. Rain splashed against the windows. It was pouring in Washington. Buck stared at gray buildings outlined against the gray sky. Washington looked very different in the rain. A few weeks earlier, he had been in Washington on the hottest day of the summer. For weeks he would talk about how he had never been that hot in his life, not even in Florida. Now, though, it was cool. Umbrellas floated along sidewalks. Taxicabs sprayed water over curbs. Buck had come to testify before Congress.

His testimony was a formality revolving around a formality. A Missouri senator, Jim Talent, submitted a bill that would grant national designation to the Negro Leagues Baseball Museum in Kansas City—in essence make it "America's Negro Leagues Baseball Museum." Everybody around the museum thought this was a very good idea, though nobody seemed exactly sure what it meant or why it would make any difference. It sounded good. Buck came out to Washington to testify on behalf of his museum. Washington insiders called this a "dog-and-pony show."

"I've never testified before Congress," Buck said happily, as if it had been high on his list. The car pulled up to the Russell Senate Office Building. Buck was told this was the oldest of the Senate office buildings. He enjoyed the trivia. Everything about this day seemed exciting to him, and new. He followed aides through several corridors and hallways until we reached the office of Jim Talent. From there, he was taken to a room. A glass lectern was set up in the middle of the room. Jim Talent said a few words to the people seated there, and then Buck was called up. He spoke about how wonderful it was to play in the Negro Leagues, and how important it was to him that the memories stayed alive.

And then he told a story—a rare story—about his wife of fifty-one years, Ora Lee Owens.

Buck hardly ever talked about Ora. He spoke often about how they met, in Memphis, Tennessee, Easter Sunday of 1943, but after that she disappeared from his narrative. Sometimes, when among friends, he would say she was loyal and loving, and she hated his driving. Ora's faults had faded in his memory since her death, her quirks too. Once, on a long plane ride, Buck talked about how much he loved children. I asked him why they never had any. He would usually joke about childlessness ("We had a lot of fun trying" was the usual punch line), but on that day, thirty-two thousand feet above the ground, he grimaced and said, "I was never around. I loved Ora very much. But she knew that baseball was my life." In his voice you could hear the echoes of late-night arguments and long-distance phone calls. Buck broke the long silence by asking about a young Kansas City Royals pitcher. Ora did not come up again.

In this room in Washington, Buck talked about the day the new Negro Leagues Museum opened. That was 1997. Ora Lee was dying; cancer had ravaged her. She kept telling Buck that she just wanted to see the museum open. It was, in so many ways, what their lives had been about. "Our child," Buck said. On the day of the opening, Ora was still living, but only barely. She could not make it out. She told Buck to go. And he did.

That evening there was a party to celebrate the opening. The next day Buck was at a baseball-card show when he had a feeling. He excused himself and went to spend those hours with Ora.

"She said, 'I made it,'" Buck said. "And she died in my arms."

Some of the people sitting in chairs let tears roll down their faces. They asked Buck a few questions about what this national designation would mean, and Buck answered as best he could considering that he had no idea. He signed a few autographs and talked to reporters afterward. When it ended, he hugged Jim Talent and said, "That wasn't so hard. Testifying before Congress."

"That wasn't Congress, Buck," a friend said. "That was a press conference. You don't testify until later."

Buck smiled, and his face blushed crimson. He said, "Oh. Okay. Well, it still wasn't so hard."

To Buck, Washington seemed in constant motion— like a baseball field during a triple. Aides rushed this way, senators rushed that way, nobody seemed to have an instant to waste. The pace fascinated Buck for a while, but then it made him dizzy and woozy, the way people feel on that third day in Las Vegas. "How come people run around here so much and yet all anybody says is that nothing gets done?" he asked Jim Talent.

"You just touched on the Great American Question, Buck," Talent said.

Buck found himself pushed and pulled through a maze of hallways and a series of elevators. Buck had absolutely no idea where he was going or where he had come from, and he did not want to ask. Here was Hillary Clinton's office. There was a famous statue. Here was where some bill was conceived. There was a little television studio. Here was an underground tunnel. Buck felt like the children in Willy Wonka's chocolate factory. He was taken from strange room to strange room, and felt too awed and scared and eager to ask what would happen next. After a walk, Buck was helped onto a little subway. His curiosity got him.

"What's this?" Buck asked.

"This is a train to the Capitol."

"Whee!" Buck yelled, and the subway traveled about five hundred yards.

BUCK COULD SEE Washington's stars and power brokers in every direction as we sat down in the Senate Dining Room. He did not know who they were, but he knew they were stars. Samuel Alito, nominee for the Supreme Court, ate a sandwich at this table, while New Mexico's Senator Pete Domenici had the salad at that one. On a whisper, Buck turned left to see Pennsylvania senator Arlen Specter—that may have been the first time anyone in Washington had found Specter by looking left. The other Pennsylvania senator, Rick Santorum, walked over to shake Buck's hand.

"Look at me," Buck said. "Senator O'Neil."

He ordered the Senate bean soup after being told it was famous. He ordered light because the woman serving said to save room for the peach cobbler. He talked about steroids and baseball, the struggles of his Kansas City Royals and the sadness he felt because so many kids had stopped playing baseball. Then he said the peach cobbler was fabulous. Jim Talent excused himself, walked away, and returned barely a minute later. "Quick bathroom break," Buck said.

"No, that wasn't a bathroom break," Talent said. "I had to go vote."

"You mean, you just went up and voted on something?"

"Yes," Talent said, and he pointed up to a clock that showed when the next vote was to take place. Buck shook his head.

"I sure wish my mother had lived to see me here," he said.

BUCK HAD NOT asked anybody what he was supposed to say at the hearing. He expected to talk about his love of baseball and the Negro Leagues, and nobody told him any different. A statement on his behalf was submitted to the United States Senate Committee on Energy and Natural Resources Subcommittee on National Parks. Buck never saw it.

Because this bill went through the subcommittee that oversaw national parks, there were others there to make cases for their various causes. A man talked about the Yuma Crossing National Heritage Area Act. A woman made a passionate plea to get more money for sites honoring past U.S. presidents. She said many of these sites were in desperate shape. She said the Chester A. Arthur site, in particular, needed massive renovating. Senator Daniel Akaka from Hawaii, who chaired the committee that day, hardly seemed impressed. Someone in the back whispered that Chester A. Arthur himself had needed massive renovation.

Then it was time for Buck to speak. In his statement, the

one he never saw, he told the committee that his name was John Jordan O'Neil but most people called him Buck. He was the grandson of a slave, and that was where he got the name O'Neil. He knew his grandfather Julius, who had worked the cotton fields in a stretch of the Carolinas he never tried to find when he got old. Julius was a proud man. He still had faith. That was what stuck with young Buck. His grandfather still believed that the world was good.

In the statement, Buck talked about those men who had played baseball before Jackie Robinson, and how they played a free-form style of baseball—something close to jazz. The men stole third, stole home, bunted, stretched singles into doubles, doubles into triples. There was always movement on the field, movement and talk, taunts, jokes, revenge, laughter, threats, challenges. Some games were played under portable lights that sounded like circular saws buzzing. On Easter Sunday, men wore new straw boaters to the games, and on Labor Day they threw those boaters on the field after the final out. The players, according to Buck's statement, dressed and drove in style. "We played good ball, entertained crowds, fed our families, and proudly lived our separate lives," it read.

In the statement, Buck talked about the times when the players could not find a place to eat or sleep. He talked about separate bathrooms and drinking fountains. There was no high school for Buck O'Neil to attend, the biggest regret of his life. "During my ninety-four years I have learned a lot," the statement read. "But most importantly I have learned that love and education heal all wounds." In the statement, he asked for the museum to be given its national designation. He called this "critical to our ability to preserve and display this important time in American history."

But, as you know, Buck never saw the statement.

"I'm ninety-four years old," Buck said to three members of the committee. "Good black don't crack."

THERE WERE ACTUALLY only two members of the committee there to hear Buck. A third senator—Ohio's Mike DeWine—had come in. "It's wonderful to have the Honorable Mike DeWine with us," Senator Akaka said.

"I'm here to listen to Buck O'Neil," DeWine said.

Buck told them that he had been a lot of places and he had done a lot of things. He hit for the cycle. He made a hole in one in golf. He shook hands with President Truman. He shook hands with President Clinton. "And I hugged Hillary," he said.

"But I'd rather be right here, right now, than anyplace I've ever been in my whole life," he said. "This right here is one of the proudest days of my life."

Then, in words very different from the statement, he spoke about the Negro Leagues and how men denied their place in America created their own America. "This is the greatest country in the world," Buck said. "I've been all over the world. But you can't beat the US of A. Here you can be whatever you want to be."

He paused and looked around the room.

"I'm living proof of that," Buck said softly.

He told the committee members that Negro Leaguers could really play ball. He told them it wasn't a sad time. "We overcame, see," he said. "That's the lesson of the Negro Leagues." When Buck finished speaking, people from all over the Capitol rushed over to get a handshake, an autograph, a photo. Buck turned and smiled and asked, "How'd I do?"

"That was amazing," Senator Talent said. Buck shook his head.

"There was something else I wanted to say," he said, "but I couldn't figure out exactly how to say it."

BUCK HAD TO sit down. The Washington rush finally overtook him. Aides rushed over and asked if he wanted water. Buck said no. He just needed a few minutes of calm. Someone asked if he wanted to go see Hillary Clinton, who was just down the hall. Buck shook his head. "I'll see her next time I testify before Congress," he said. "Let's go home."

Out in the halls, people of various ages wore black. Lights were on in the offices, though it was still the afternoon—the skies had grown dark and raindrops slapped against the windows. Two men stood in a corner and talked in whispers, the iconic image of backroom Washington dealings. Buck closed his eyes and tried to shut it all out. After a few seconds, he was told the car had arrived, and everyone stood up.

"Hold on for a second, hold on," Buck said, and he pointed at a television nearby. "You know something funny? Look at that television. You know, if the Willie Mays catch was on right now—the one from the World Series—everyone would stop and watch it."

Everybody around stopped and listened to Buck. He was talking about the catch Mays made at the Polo Grounds in the 1954 World Series, the one where he turned his back and raced toward the wall on a long ball hit by Vic Wertz. Mays ran full speed to a spot and somehow caught the ball over his head without even looking back. His hat flew off. And then, in one motion, Mays whirled and threw the ball back to the infield. Jack Brickhouse, the announcer, screamed that it must have looked like an optical illusion to a lot of people. More than fifty years later, most people would say it was the greatest catch ever made.

"How many times have we all seen that catch?" Buck asked. "And yet, if Willie Mays was up there on the television, this whole place would come to a stop."

> *If Willie was up there*
> *People would stop making laws.*
> *They would stop running.*
> *They would stop arguing about*
> *Little things*
> *Or big things.*
> *No Democrat or Republican,*
> *No black and white,*
> *No North or South.*
> *Everyone would just stop,*
> *Watch the TV,*
> *Watch Willie Mays make that catch.*
> *That's baseball, man.*

Buck smiled, shook some hands, and walked out to the car. After all this time, he had come to Washington and said what he wanted to say.

WINTER
(TAKE 2)

HOME

Buck O'Neil jolted awake when the telephone rang. He was home in Kansas City. The voice on the other end claimed to be a radio producer and he asked if Buck could do a quick interview. Buck could not remember saying yes, but in the next instant, there was a radio host on the line with that familiar deep voice Buck had heard thousands of times in every city in America. The radio voice asked him if he was excited about going into the Hall of Fame.

"Well, it hasn't happened yet," Buck said.

"What do you think your reaction will be when you get the word?"

"Excited," Buck said. "I'm sure I'll be excited."

Buck got out of bed. He walked on the treadmill for a few moments, though his heart was already beating fast. Buck thought about what to wear—this was always a big decision. He decided to go informal. He put on a pair of jeans, a Kansas City Monarchs jersey, a Kansas City Monarchs bomber jacket, and a Negro Leagues Baseball Museum hat. "This is baseball," he said.

He showed up at the Negro Leagues Baseball Museum at 10 A.M. sharp. He asked Bob Kendrick, "When will word come?"

"About eleven A.M."

"All right, then, I'm going to make a few phone calls," Buck said. He reached into his wallet and shuffled through dozens of business cards and scraps of paper. He found a number and dialed it on a white phone nearby. It took him two tries to dial it right. He waited a long time for someone to answer, and when someone did, he jumped right into the conversation: "It's Buck. If the vote goes my way, I'll be coming to see you. Yeah. . . . I've missed you so much too. . . . Yeah. Come a long way from those days. Ha ha ha. . . . Yeah."

He hung up. He said that was his cousin. She was ninety-three, one year younger than he was. And years ago, when they were both young in Sarasota, she used to play catch with him.

NEWSPAPER STORIES OFTEN follow a theme. Most of the stories leading into this day would have fit under the headline BUCK O'NEIL AND OTHERS TO ENTER BASEBALL HALL OF FAME. Buck had involuntarily overpowered the day. The Baseball Hall of Fame decided to right some wrongs and induct some of the great Negro Leagues players and contributors who had been overlooked. They put together a panel of twelve Negro Leagues experts—academics, writers, historians—and gave them free rein. Elect at will, the Hall of Fame said. Make things right.

The panel put together a ballot of thirty-nine players, managers, and executives from the Negro Leagues. The ballot burst with stories—there were many great players, shady characters, heartbroken men, a single woman—but focus turned to Buck O'Neil.

"You know why that is," Buck said. "Because I'm alive."

This was true. Only two men on the ballot—Buck and Minnie Minoso—were still living. But there was something else too. For so many years, Buck had traveled the country talking about the Negro Leagues. He had done countless interviews. He had appeared on every rowdy morning show and spoken in every whispered Negro Leagues documentary. He had told his stories to David Letterman and Ken Burns and Jim Rome. He had been quoted in virtually every obituary of every man who had played in the Negro Leagues. Buck had, in fact, become the Negro Leagues to millions of Americans. He kept the Negro Leagues' reality fresh and vibrant. Yes, sure, it would be nice to get a few deceased men their rightful place in the Hall of Fame. Sonny Brown, the man who played breathtaking baseball in the sunshine, was on the ballot. Biz Mackey, who old-timers said was a better catcher than anyone who ever lived before or since, was on the list. There were others who were worthy. They were dead.

Buck was alive. This day was about Buck O'Neil. And everyone knew that, even Buck.

THE MUSEUM PHONE did not ring at 11 A.M., and Buck settled in for a long wait. Buck said he had been on so many committees like this one. He said things never go quite like you expect. He did a television interview. He said that if the Hall of Fame call came, it would be one of the great moments of his life. "This is what every ballplayer dreams of," he said. "I'm no different than anyone else when it comes to the Hall of Fame. I'm in awe."

People wandered into the room to hug Buck and wish him luck. Lou Brock, one of the Hall of Famers Buck had signed as a scout, called. "I'm excited, Buck," he said.

"Me too," Buck said.

"You know I'm there for you."

"Yes I do, Lou. Yes I do."

"There are angels everywhere, Buck," Lou said.

Buck smiled. That line went back to a story. When Lou Brock tried out for the college baseball team at Southern University, he felt scared to death. Brock knew he could run, but he did not believe he could hit against college pitchers. He felt sick at the tryouts. Everybody looked bigger than him, stronger than him. They all looked surer than him. He wanted to run away. Then he saw a young boy on the field. The boy could not have been more than twelve years old, but he acted as if being there and playing baseball with the giants was as natural as running with the wind. He played catch with everybody. He shouted out. He ran the bases as if he belonged.

Lou Brock thought: *If a little boy has the courage to run on this field, so do I.* Lou hit a long home run in those tryouts. He ran free. His future was in motion. In time, he would steal more bases than any Major League player. He would crack three thousand hits. And Lou Brock was at his best in the biggest games—Brock played in three World Series and he was a dominant presence in all three. He hit .391 in those games and stole fourteen bases. He felt sure.

Anyway, after the tryout, Brock wanted to thank the little boy for helping him overcome his fears. But when the tryout ended, he looked everywhere and could not find the boy. He went up to his new teammates and said, "What was that little boy's name?"

And they said: "What little boy?"

"You know, the one that was running around here, playing catch with everybody."

And they said: "There was no little boy out here."

Lou Brock said he never saw that boy again. One time he told that story, Buck O'Neil was sitting next to him. Buck asked: "You think that boy was an angel?" Brock smiled.

"There are angels everywhere," Buck said.

MINUTES TICKED SLOWLY by, as if all the clocks in the room were underwater. Buck tried to make time pass by talking about the events of the day. The Winter Olympics had just ended. The president's approval ratings were down. The weathermen on television talked snow. Bob Kendrick tapped me on the shoulder and asked me to step outside for a moment.

"It's not looking good," Bob said. His expression locked halfway between enraged and hopeless.

"What do you mean?"

"I just got a call. They said that the committee had some sort of straw poll just to see where everything stands. And Buck is short some votes."

"How is that possible?"

"I don't know. I hear that former commissioner Fay Vincent has asked everyone on the committee to strongly consider how much Buck has done for baseball. But I don't know if it's going to make much of a difference. It doesn't look good."

Through the window of the room, the sky looked darker as Buck talked about how crazy bobsledders must be. He looked over, caught Bob's eye, and kept on talking. "They just get in that sled and go a hundred miles an hour on ice? Sorry. Give me a bat and a ball."

"I don't know how I'm going to tell Buck if he doesn't make it," Bob said.

"He will make it," I said.

"I hope so."

Time crawled. There was too much to think about. How was it possible that this panel would not elect Buck O'Neil into the Hall of Fame? I played it over in my mind. Buck was not a great player. Maybe that was the problem. He had been a good player—he led the Negro Leagues in hitting once, almost did it again the next year, and he had prime years taken away by the war—but he was not a great player. Buck had always said he was not a Hall of Fame player, if that was the whole story.

But, of course, his playing was not the whole story, not even half of it. Keeping Buck out of the Baseball Hall of Fame because he was not a good enough player, I thought, would have been like keeping Leonardo da Vinci out of the Renaissance Hall of Fame because he was not a good enough inventor. Buck had been a great manager. He led the Monarchs to pennants. He sent more Negro Leagues players to the Major Leagues than anyone else. He would have been the first black manager in the Major Leagues, surely, but the Major Leagues were not ready for a black manager. Buck was, instead, the first African-American coach in the big leagues.

Buck was a great scout too. Scouts for some reason don't have their place in the Baseball Hall of Fame, but how could they ignore that as part of Buck O'Neil's amazing life? And all of those were just appetizers—he was a spokesman for the game. The greatest spokesman. That was his legacy. He was the living voice of the Negro Leagues for more than fifty years. Could this not matter? For all those years—even now, at ninety-four—Buck traveled the country promoting baseball, selling it, teaching it to kids. How could they not vote him into the Baseball Hall of Fame?

As the minutes ticked slowly by and the clock approached noon, Buck seemed to understand what was happening. He said suddenly and with force: "You know something? I could play. I was no Josh Gibson, but I could play." His face, for an instant, looked pained. It was the closest I had seen to Buck cracking. Bob and I looked at each other. And just like that, Buck's face slackened, and his smile reemerged, and he talked again about everyday things.

ANOTHER HALF HOUR passed, and everyone in the museum began to understand and recognize the harsh truth. Buck said, "I'll be all right either way, really." Nobody believed him entirely. The Hall of Fame does not make a man's career, of course. It is only a museum in a city where baseball did not begin. There were men inducted because they were born wealthy enough to buy baseball teams. There were others, deeply flawed men, inducted, though they tainted the game as much as they celebrated it. Cap Anson was a great player. He was also a virulent racist. He, more than any other man, drew the color line. Gaylord Perry was a great pitcher. He spit on baseballs and cheated to win. Ty Cobb attacked a fan. And so on.

So, no, the Hall of Fame would not define Buck O'Neil, but as he sat there in that dark room, and he stared at that slow-moving clock, it was clear that he wanted it. That was the heartbreaking part. Buck had not wanted much. He had suppressed his bitterness about missing out on the Major Leagues as a player and a manager. He had put away his anger over being ignored and tuned out. Buck had worked hard to get the players he admired into the Hall of Fame—all the time refusing to ever campaign for himself. Now, in that room, he had hoped. And he slowly began to realize that they were going to pass him by again.

At 12:34, Bob walked into the room and closed the door. His eyes were bloodred. He said: "Buck, we didn't get enough votes."

Buck O'Neil said: "Well, that's the way the cookie crumbles." The room was quiet for a couple of moments. Bob said he did not have any details yet, but he heard seventeen people had been inducted.

"Seventeen!" I gasped. I didn't mean to do it. My voice had just escaped, the way a groan escapes when a long fly ball hooks foul at the last moment. Seventeen! There were only eighteen Negro Leaguers in the Hall of Fame. It did not seem real. They had inducted seventeen people into the Hall of Fame, and Buck O'Neil had not been one of them? That seemed impossible. A cruel joke. Buck looked over at me. His eyes danced.

"Seventeen, huh?" he asked. "That's wonderful."

THERE WERE, IN fact, seventeen people voted into the Hall of Fame that day. One of them was the first woman ever inducted, Effa Manley, who had co-owned the Newark Eagles. Sonny Brown had made it, which delighted Buck no end. Biz Mackey had made it. Bob Kendrick walked into the Field of Legends in the heart of the museum to tell the press that Buck had not made the Hall of Fame. Bob began to speak, and then he broke down and started sobbing. Some people ranted in the hallways. This was Buck's home field, and righteous indignation ignited the air. They wanted to know how this was possible. They would not find out. The committee members would not explain why Buck was not voted into the Hall of Fame. They had decided to keep their votes secret.

Back in that little room with the white phone, Buck looked at the list of new Hall of Famers for a long time. He did not say much, but he nodded sometimes and his eyes flickered recognition as they scanned the names. He did not look any older or younger than he had a few minutes before, when he was talking about bobsledders and how much he loved baseball. He did not look any happier or sadder. I asked him how he was feeling. He said:

> *I'm all right.*
> *Disappointed, maybe.*
> *I thought I would . . .*
> *Happens in life.*
> *You keep on going.*
> *Happens in life.*

He went back to looking at the names. Every so often he would call out a name and say, "I'm glad he got in. He was a great player. I'm so glad."

BEFORE HE WENT to speak to the gathered media, Buck turned to me and said: "I wonder who they will ask to speak at the Hall of Fame induction."

It was a good question. The seventeen people selected were

dead. Most of them had been gone for fifty years. It seemed unlikely that the Hall of Fame would find family members for all of them, and they certainly would not let each family member give a speech—that would be like "Open Mike Night" and the ceremony would last three days. There was word that Jackie Robinson's daughter Sharon would speak, a nice touch, but she was not old enough to know most of these players and executives. Double Duty was gone. Maybe Monte Irvin would speak?

"I wonder if they'll ask me to speak," Buck said.

I was about to laugh. But I looked at Buck. He seemed serious. He was serious. I had seen that look. He honestly wondered if the Hall of Fame—shortly after their committee snubbed him—would ask him to speak. More than that, he looked hopeful. I looked hard at the man I had been traveling with for more than a year, and I realized that even after all this time I had not learned the biggest lesson. I seethed with anger for Buck. I wanted to interrogate each and every member of that committee, shine a World War II–era lamp in their eyes and ask them if their hearts were still beating. I wanted to know how they could look at the baseball life this man had lived and not vote him into the Hall of Fame. They had put in seventeen people, none of whom had one-tenth the impact this man had. Yet when I looked at Buck in that moment, the anger seeped out of me.

"Would you really speak?" I asked Buck.

"Of course," he said happily. "If they asked me. . . . It would give me a chance to talk about these great players. It would give me a chance to tell the story again."

"Buck," I said slowly, "I don't see how you—"

He put his arm on my shoulder and said: "Think about

this, son. What is my life about?" And he went into the other room to answer questions from reporters and fans.

THERE WAS ANGER across America in the days that followed. The committee members held true to their word and did not discuss Buck O'Neil. Rumors filled the silence. One rumor went that some members, while they admired Buck O'Neil, felt he was not a good enough player to get into the Hall of Fame. In this version, it took great courage for them to refuse him induction. They were guarding the hallowed halls. Another rumor claimed the snubbing had to do with a political fight between committee members and the Negro Leagues Baseball Museum, and Buck was the collateral damage. Still another rumor had members of the committee jealous of Buck's fame. None of the rumors satisfied. They never do. Nobody stepped from the fog to explain.

Rage swept through the baseball community. The *New York Times* wrote a story asking how Mr. O'Neil could have been passed over. The *New York Post,* as usual, was more direct: "They left Buck O'Neil off the list . . . which makes the list a complete joke," *Post* columnist Mike Vaccaro wrote. More than fifty major newspapers and web sites wrote searing editorials about Buck O'Neil's absence from the list.

"He is the greatest ambassador the Negro Leagues have ever known," King Kaufman wrote for Salon.com.

"The committee should be ashamed of itself," the *Detroit News* editorial board wrote.

Players were outraged. Hank Aaron, the all-time home-run leader, called up the Negro Leagues Baseball Museum to ask what could be done. Ernie Banks cried that Buck deserved to be in the Hall of Fame more than anyone else, and he included himself. Joe Morgan, a Hall of Famer and the vice chairman of the Hall, called it a disgrace. Another Hall of Famer, Bob Feller—never one to mince words—said of the committee: "What the hell do they know about baseball?" When baseball commissioner Bud Selig first heard the news, he asked his spokesman Rich Levin: "How is this possible? How can this be? How is this even possible?"

Politicians spoke up from both sides of the aisle. Missouri Republican senator Jim Talent asked Major League Baseball to correct the mistake while Missouri Democratic congressman Emanuel Cleaver called the snub a "shameful error." The Negro Leagues Baseball Museum was flooded with angry calls. Hillary Clinton said something needed to be done. Two Internet petitions to get Buck into the Hall of Fame were started. On MSNBC, Keith Olbermann raged—he said this was the biggest mistake the Hall of Fame ever made. In the *Sporting News*, Dave Kindred raged. "I intend to get Buck O'Neil into the Hall of Fame if it's the last thing I do," he said.

And Buck? For a while, he did interviews and talked about the decision. After a while, though, he saw no use in it. The people wanted him to be angry, to hate this committee, and he simply could not do that. *Think about this, son. What is my life all about?* He said the committee had voted with their hearts. He said that he felt worthy of the Hall of Fame, but there were excellent players not in the Hall and excellent men too. Of course, the better he took the rejection, the worse the committee looked. I suspect Buck knew that. That's why, at some point, he decided not to talk about it anymore. He never turned down an interview, of course. But he shut down when they asked him about the Hall of Fame.

"You know why I wanted to get into the Hall of Fame?" he had said in the moments after the rejection. "Because if I got in, I could have fought for some of the people who still deserve to get in. Dick Lundy. Great shortstop. He should get in. Ted Strong should get in. Bill Wright should get in. But now I can't fight for them, because if I do that, people will say, 'Oh, Ol' Buck is trying to get himself into the Hall of Fame.'"

Buck shook his head. A fierce pride burned in his face. Buck wanted to go to the Hall of Fame, sure he did, but he wanted to make this point clear: He would never campaign for himself. Not ever. And he did not talk about that again.

On a hot day in Cooperstown in July of 2006—the day sixteen men and one woman from the Negro Leagues were inducted into the Baseball Hall of Fame—Buck O'Neil spoke. His voice sounded scratchy. He asked everyone in the crowd to hold hands. He asked them to sing with him. And he sang: "The greatest thing in all my life is loving you."

I was walking through a bookstore, marveling at how proudly people buy books for idiots and dummies, when the cell phone rang. It was past 10 P.M. It was Buck O'Neil.

"I want you to do me a favor," he said.

"Sure, Buck."

"I want you to thank all the people. All the people who have said nice things since the Hall of Fame thing happened."

"Of course, Buck."

"I want you to thank them for all their support. I want you to tell them that the greatest thing in all my life is their love for me and my love for them in return. Can you do that for me?"

"Of course. Absolutely."

"I have never felt more loved. All my life. Tell them that."

There was a pause then, and I figured Buck had finished talking. But the pause grew longer, and I asked Buck how he was doing. He said he was doing great. He said again that he wanted to go into the Hall of Fame, but he figured that God works in his ways, and that by not getting in, he found out just how beloved he was.

I said, "Buck, you're about the only person I know who could feel that way."

He laughed and said: "You might be right about that." Another pause.

"You know," he said, "a few weeks ago a guy asked me: 'Who is that white boy who is following you around all the time?'"

"What did you say?"

"I told him, 'Can't you tell? That's my son.'"

Buck laughed. There was another silence and I started to say something, but Buck cut me off fast. He said, "See you at the ballpark." And he quietly hung up the phone.

ONE NIGHT LATER, Buck was given an honorary doctorate degree by William Jewell College in Kansas City. It was a big event. Ken Burns was there. All night, people charged Buck to tell him how unjust it was, him not getting into the Hall of Fame, and Buck just patted their shoulders and said, "It's all right, really, it's all right." It was odd in a way, watching Buck comfort others over the injustices he endured. And then it hit me that it was not odd at all. This was exactly what Buck O'Neil had been doing all through our seasons together. And I remembered one of the first things Buck ever said to me. He said that people did not really listen to him talk about the Negro Leagues. Oh, they liked his stories. They liked him. But they already had it in their minds how it was—baseball clowning and bounced checks, men shouting racist slurs and white kids throwing rocks at their rickety buses. They needed to believe in something simple.

People used to tell me
How they thought it was
Way back then.
Used to tell me
How they imagined it.
And I tried to say
It wasn't like that.
We were men
Flesh and blood
And we played baseball in the sunshine.
We hit doubles off the wall,
Slid hard into second base.
We had fights, and we made love.
We sang songs and prayed on Sundays.
Before games.
We were real. Yeah. We laughed and cried.
We felt pain. And we felt joy.
There was a lot wrong with the world.
But we weren't sad, man.
We had the times of our lives.
I told them that for fifty years.
They heard. But they didn't listen.
They listened. But they didn't hear.

Now, though, people did listen. They heard. And in the end, he said, that was what mattered. That created the joy in his life. "Good black don't crack," he told the people as he was introduced to the crowd as Dr. Buck O'Neil. His friend Ken Burns talked about the possibilities of America. He spoke about the Civil War and the Brooklyn Bridge. He talked about the pain and joy found in jazz, and most of all, he talked about baseball and how it defines our troubles and our triumphs. In the middle of this, though, without warning, he veered away from his well-patterned words and turned toward Buck O'Neil. He said: "All we can say to him tonight is he belongs in our Hall of Fame."

Everyone in the room stood and cheered for a very long time. They would not stop. Buck O'Neil sat in the middle of those cheers. He soaked them in. His eyes were closed and he raised his arms above his head as if waiting to catch a baseball from heaven.

AFTERWORD

My wife and I bought a piano at a sale held in a college chapel on October 6, 2006. It was black and so shiny you could see your face in the wood. I cannot say the purchase was inspired entirely by Buck O'Neil. But I can say I heard his voice in my mind when I hit the keys with one finger and listened to the tones.

"Son," Buck had said in New York, "in this life, you never walk by a red dress."

I think Buck meant that we should never pass up the opportunity to live life. We should not rush by the red dresses, the baseball games, the street musicians, or the sweet smell of dessert. We should not stifle or smother our craziest dreams. I had always wanted to play the piano, so we bought it, and they delivered it to the house that day. I was playing some sort of off-key jazz thing that night when the phone rang. I kept playing. The phone rang again and still I kept playing. The phone rang a third time, and I knew.

Buck had been in the hospital for a few weeks by then. When I saw him there, he had lost a lot of weight, and he did not have a lot of weight to lose. He had stayed trim through the years on the Buck O'Neil diet. "Two meals a day," he used to say. "That's all you need." In the hospital, friends had to beg him to eat at all.

He had also lost his beautiful voice. Doctors could not explain why. His deep baritone had dissolved into a raspy whisper, and I think that bothered him more than the bone-aching fatigue that had overtaken him. Buck felt naked without his voice. When he saw me, he had to pull me close. He whispered in my ear: "You are my friend."

"Of course, Buck."

"I hear the book is done," he whispered. "How is it?"

I promised him I would bring him a copy. We sat close and talked for a while. His memory was still sharp. He talked about how much he loved Cajun food, he remembered some great Negro Leagues players, he said that kids today face dangers that children of his time never knew. "I wish they would stop killing each other," he said.

Buck also said he wasn't quite ready to die yet. He still had things he wanted to do. Then he coughed something fierce. I didn't know that that would be the last time I would see him. You never know.

"I'm going to bring that book by," I told him as I was leaving.

"Anytime," he said. "Anytime."

Three days later his health took another bad turn. He stopped having visitors. "I don't think Buck would want you to see him like this," our friend Bob Kendrick said. "I think he would want you to remember him the way he was when we traveled around America."

Buck lasted a week longer than friends and doctors expected. Buck O'Neil died that October night I was trying to play jazz on a shiny new black piano. Baseball and jazz, he had always said, were the two best things in the world. Of course, I was just plinking keys on the piano. I wasn't really playing jazz.

"It's all jazz," Buck had said.

Buck was ninety-four years old, almost ninety-five. He had asked me not to cry when he died, but I did anyway.

ACKNOWLEDGMENTS

I'm a compulsive acknowledgments-page reader, probably because I love to see who people know. Of course, it now occurs to me you don't have to actually know the people you acknowledge, and so I want to thank Bruce Springsteen, Nick Hornby, Jerry Seinfeld, Natalie Portman, Jon Stewart, Chris Rock, Meryl Streep, Ben Folds, and numerous other cool people who may or may not have had any actual impact on this book.

Thanks again to Mike Vaccaro, Jim Banks, Richard Bush, Ken Burger, Brian Hay, Michael MacCambridge, Ian O'Connor, Tommy Tomlinson, Adrian Wojnarowski, and Bill James for helping shape this book in different ways. The good stuff in here is theirs. Thanks to the *Kansas City Star*, and in particular my friends and bosses Mike Fannin, Mark Zieman, and Holly Lawton, who encouraged me through write the book even if it meant missing a few newspaper columns. Thanks to my literary agent and publisher. Sloan Harris made this book possible. David Highfill made it better. William Morrow & Company made it.

Thanks to my in-laws, Cecil and Judy Keller, for their babysitting and love during the book crunch. Thanks to my own parents, Frances and Steven, and my brothers Tony and David. Thanks especially to my daughters, Elizabeth and Katie, for keeping me sane by attacking me with very small dolls.

So many people helped during the year I spent with Buck, and I will undoubtedly forget some. Here's my best effort: Thank you to the good people at Roadway and rEvolution, Annie Presley-Selanders, numerous Major League baseball teams, Bob Turnauckas, Tony Oliva, Mudcat Grant, Willie Mays, Monte Irvin, Dave Winfield, Hank Aaron, Jimmy Wynn, Ernie Banks, Lou Brock, Luis Tiant, Joe Delamielleure, Billy Williams, Minnie Minoso, J. C. Hartman, and countless former Negro Leaguers who generously shared stories and feelings. Most of all I thank Don Motley, Ray Doswell, and the people at the Negro Leagues Baseball Museum in Kansas City. Some proceeds from this book will go to the museum, and I cannot recommend strongly enough that you see this magical place for yourself. Check it out at www.nlbm.com.

There are two acknowledgments that must be set apart. First, I want to thank my friend Bob Kendrick, who went along for most of this trip and helped make it possible. This is his book as much as it is mine. And second, I want to thank my love, my best friend, and my first editor, Margo, who said a few years ago, "You need to write a book about Buck O'Neil," and then patiently and lovingly suffered the consequences of those words. This book is dedicated to her. I hope this entire book serves as a thank-you to Buck O'Neil.